The SpringerBriefs Series in Materials presents highly relevant, concise monographs on a wide range of topics covering fundamental advances and new applications in the field. Areas of interest include topical information on innovative, structural and functional materials and composites as well as fundamental principles, physical properties, materials theory and design. SpringerBriefs present succinct summaries of cutting-edge research and practical applications across a wide spectrum of fields. Featuring compact volumes of 50 to 125 pages, the series covers a range of content from professional to academic. Typical topics might include:

- A timely report of state-of-the-art analytical techniques
- A bridge between new research results, as published in journal articles, and a contextual literature review
- A snapshot of a hot or emerging topic
- An in-depth case study or clinical example
- A presentation of core concepts that students must understand in order to make independent contributions

Briefs are characterized by fast, global electronic dissemination, standard publishing contracts, standardized manuscript preparation and formatting guidelines, and expedited production schedules.

More information about this series at http://www.springer.com/series/10111

Martin W. King • Tushar Bambharoliya
Harshini Ramakrishna • Fan Zhang

Coronary Artery Disease and The Evolution of Angioplasty Devices

Springer

Martin W. King
Wilson College of Textiles
North Carolina State University
Raleigh, NC, USA

College of Textiles
Donghua University
Shanghai, China

Harshini Ramakrishna
Wilson College of Textiles
North Carolina State University
Raleigh, NC, USA

Tushar Bambharoliya
Wilson College of Textiles
North Carolina State University
Raleigh, NC, USA

Fan Zhang
Wilson College of Textiles
North Carolina State University
Raleigh, NC, USA

ISSN 2192-1091   ISSN 2192-1105   (electronic)
SpringerBriefs in Materials
ISBN 978-3-030-42442-8   ISBN 978-3-030-42443-5   (eBook)
https://doi.org/10.1007/978-3-030-42443-5

This Springer imprint is published by the registered company Springer Nature Switzerland AG
The registered company address is: Gewerbestrasse 11, 6330 Cham, Switzerland

Abstract

Coronary artery disease (CAD) is a major cause of death and disability in the world. Heart disease and strokes are among the top five leading causes of death. In 2010, ischemic heart disease was the cause of 13.3% of all deaths globally. The 2016 Heart Disease and Stroke Statistics update of the American Heart Association reports that 15.5 million adults (6.2% of the adult population) in the United States have CAD, including 7.6 million (2.8%) with myocardial infarction (MI) and 8.2 million (3.3%) with angina pectoris. By 2035, nearly half of the US population (45.1%) will have some form of cardiovascular disease. The direct medical costs from coronary artery disease and stroke are currently estimated to be $126 billion per year and are expected to rise to $309 billion by 2035.

In this book we have attempted to provide the current issues related to coronary artery disease and take the reader through the epidemiology, risk factors, and the basic mechanism of coronary artery disease. In addition, we explain how coronary microvascular dysfunction can lead to angina, even after successful angioplasty following a heart attack. The methods for diagnosis and the options for treatment are explained, and we end with a discussion of the evolution of coronary angioplasty from balloon catheters to novel, fully resorbable polymeric drug-eluting stents. We have enclosed a list of currently available commercial devices, as well as a listing of experimental products under research and development, so that readers can appreciate the serious nature of this disease and have an understanding of how the medical device sector is attempting to address this major healthcare problem.

Contents

List of Figures

List of Tables

Chapter 1
Epidemiology and Risk Factors

Keywords Epidemiology · Risk factors · Life's Simple 7 · Coronary artery disease · Ischemic heart disease · Myocardial infarction · Cardiovascular diseases

Coronary artery disease (CAD) is a major cause of death and disability in developed countries. Heart disease and stroke are among the top five leading causes of death. In 2010, ischemic heart disease was the cause of 13.3% of all deaths globally [1]. Cardiovascular disease (CVD), listed as the underlying cause of death, accounts for nearly 836,546 deaths in the United States. That's about one of every three deaths in the United States according to the Heart Disease and Stroke Statistics 2018 report by the American Heart Association (AHA) [2, 3]. About 2300 Americans die of cardiovascular disease each day, an average of one death every 38 s. Coronary heart disease (CHD) (43.8%) is the leading cause of deaths attributable to cardiovascular diseases (CVDs) in the United States, followed by stroke (16.8%), high blood pressure (9.4%), heart failure (9.0%), diseases of the arteries (3.1%) and other cardiovascular diseases (CVDs) (17.9%). The 2016 Heart Disease and Stroke Statistics update of the AHA reports that 15.5 million adults (6.2% of the adult population) in the United States have CAD, including 7.6 million (2.8%) with myocardial infarction (MI) and 8.2 million (3.3%) with angina pectoris. A Global Burden of Disease Study Group report from 2013 estimated that 17.3 million deaths worldwide in 2013 were related to atherosclerotic cardiovascular disease (ASCVD), a 41% increase since 1990 [4]. By 2035, nearly half of the US population (45.1%) will have some form of cardiovascular disease [5]. Currently, approximately 16% of the overall national healthcare expenditures in the United States are used to manage the burden of CVD healthcare costs [6]. The direct medical costs from coronary artery disease and stroke are currently estimated to be $126 billion per year and are expected to rise to $309 billion by 2035. In 2016, in America total cost of CVD including direct and indirect costs were $555 billion. By 2035, the costs will skyrocket to $1.1 trillion [7].

M. W. King et al., *Coronary Artery Disease and The Evolution of Angioplasty Devices*, SpringerBriefs in Materials, https://doi.org/10.1007/978-3-030-42443-5_1

Heart Disease, Stroke and Cardiovascular Disease Risk Factors

The American Heart Association has defined seven key risk factors for heart disease and stroke called "Life's Simple 7". These risk factors are smoking, physical inactivity, nutrition, overweight/obesity, cholesterol, diabetes and high blood pressure. A recent major study among African Americans found that the risk of heart failure was 61% lower among those with equal or greater than four of these ideal cardiovascular health metrics compared to those with only zero to two positive metrics. One thing to note that when the risk factors are controlled, outcomes are improved irrespective of race [3].

There are currently 1.3 billion people in the world who use tobacco and almost one-third of coronary heart disease deaths are attributable to smoking or exposure to secondhand smoke. It is estimated that active and passive smoking remains responsible for more than 480,000 deaths annually in the United States, and, if this trend continues, tobacco will be responsible for more than 1 billion deaths during the twenty-first century. Elevated blood pressure is one of the leading cause of coronary heart disease (CHD). Worldwide, blood pressure above 115 mm Hg systolic is the reason for approximately 62% of strokes and 49% of myocardial infarctions, leading to more than 7 million deaths annually. High cholesterol contributes to around 56% of ischemic heart disease and 18% of strokes leads to 4.4 million deaths annually. Diabetes, obesity and physical inactivity are also major contributors to this disease. Diabetes, a major risk factor for CVD, currently affects 17 million Americans. Exercise can reduce around 30–50% of the risk of coronary artery disease. Physical inactivity is considered to be the fourth leading cause of death worldwide [6, 8–10].

Chapter 2
Definitions and Basic Mechanism of Coronary Artery Disease (CAD)

Keywords Atherosclerosis · Plaque · Thrombosis · Blood coagulation · Plaque rupture · Calcification · ST-elevation myocardial infarction · Non-ST-elevation myocardial infarction · Electrocardiography

Coronary artery disease (CAD) also known as coronary heart disease (CHD) is described as the pathologic process affecting the coronary arteries, while atherosclerotic cardiovascular disease (ASCVD or CVD for short) is referred to as the pathological process affecting the entire arterial circulation, not just the coronary arteries. The diagnosis of angina pectoris, myocardial infarction (MI) and silent myocardial ischemia are morbidities and events included within CAD. MI leading to death is considered an endpoint of CAD, while transient ischemic attacks, angina, MI, claudication pains, stroke and critical limb ischemia are symptoms of CVD in general. CAD can be further classified as either stable coronary artery disease or stable ischemic heart disease (SIHD), and acute coronary syndrome (ACS), which is further classified as MI and unstable angina. Stable coronary artery disease patients, who have a prior history of current demonstrable obstructive atherosclerotic disease of the epicardial coronary arteries are either asymptomatic or have stable symptoms, with no evidence of recent symptomatic, hemodynamic or electrical decomposition (ECG abnormality) [11, 12].

Atherogenesis or atherosclerosis refers to the development of atheromatous plaques associated with asymmetric focal thickening along the inner lining of the arteries. This is considered the main cause of CAD and is an inflammatory disease in which metabolic risk factors trigger immune mechanisms to initiate, propagate and activate plaque formation and lesions in the arterial tree. The wall of the coronary artery has three layers named the tunica intima, tunica media and tunica adventitia. The tunica intima layer contains the endothelial cell layer having direct contact with the blood flow. Internal elastic lamina separates the tunica intima from the tunica media layer. The tunica media is made up of a layer of circumferential smooth muscle cells (SMCs) and variable amounts of connective tissue. The external elastic lamina separates the tunica media from the tunica adventitia layer. The tunica adventitia consists mainly of connective tissue fibres and blends with the connective tissue surrounding the vessel. All three layers are shown in Fig. 2.1 [13–16].

The tunica intima endothelial cell layer that is in direct contact with the blood flow has the ability to resist deposition and attachment of white blood cells. At the same time, the expression of adhesion molecules on the endothelial cell layer attempts to capture leukocytes on the surface due to such factors as dyslipidaemia, hypertension and pro-inflammatory mediators that trigger the initiation of atherosclerosis. Subsequent endothelial layer permeability causes the extracellular matrix composition to be modified, which results in permeability changes and the influx of cholesterol-containing low-density lipoprotein (LDL) particles into the tunica intima. The oxidative and enzymatic modification of LDL in the tunica intima triggers the endothelial cells to express leukocyte adhesion molecules [14]. Such oxidized LDL activates specific pattern recognition such as toll-like receptors (TLR). These chemoattractant mediators cause the myocytes or leukocytes to migrate into the tunica intima layer as shown in Fig. 2.2 where these white blood cells differentiate into tissue macrophages [17]. The LDL particles residing in the arterial wall become engulfed by these macrophages and become known as foam cells or lipid-laden macrophages [18]. These macrophages in the plaque on the intimal layer also release other cell recruiting molecules named cytokines such as interleukin-1β (IL-1β) and tumour necrosis factor (TNF) [18, 19].

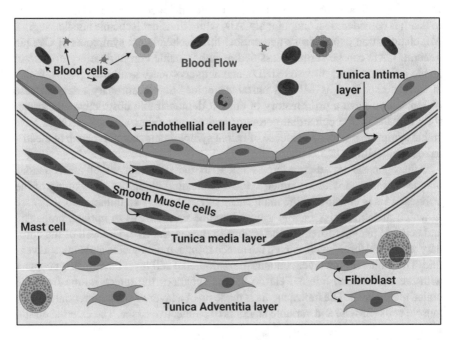

Fig. 2.1 Three layers of coronary artery wall consisting of tunica intima, media and adventitia [13, 15, 16]. (Created with BioRender.com)

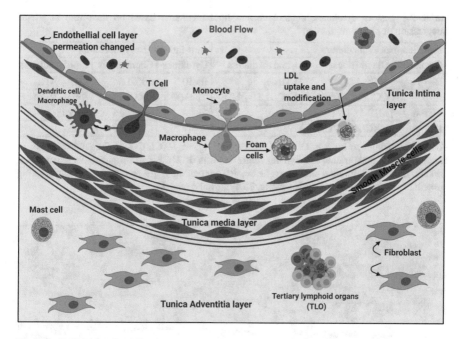

Fig. 2.2 Cellular mechanism of atherosclerosis [20–22]. (Created with BioRender.com)

The plaque or the atheroma start to attract other leukocyte cells such as lymphocytes as well as mast cell and immune T cells, and all these cells are responsible for the key regulatory function of the plaque. Smooth muscle cells (SMCs) from the tunica media layer start to migrate into the tunica intima and the plaque site and proliferate in response to mediators such as platelet derived growth factor (PDGF). These SMCs start producing extracellular matrix, including interstitial collagen, proteoglycans and elastin, that form the fibrous cap around the plaque [23, 24]. The foam cells eventually undergo apoptosis, which leads to the release of the lipid into the extracellular space of the plaque. This lipid, as well as dead cell debris, form a lipid-rich pool and necrotic tissue in the plaque sack within the tunica intima layer. Activated macrophages in the plaque region play an important role in stabilizing the atherosclerotic lesion. Table 2.1 shows the list of cytokines, leukotrienes, mediators and growth factors that either help to promote the resolution of the inflammation and ultimately promote tissue repair and healing or assist in sustaining the inflammatory response and promote tissue necrosis [20, 21].

The plaque underneath the endothelial cells is covered by a thin fibrous capsule, which does not trigger any major thrombotic complication until there is physical disruption of the thin, collagen-poor fibrous capsule which exposes the plaque to the blood vessel flow as shown in Fig. 2.3 [25, 26].

This capsule can be sustained for longer periods of time without disruption, during which the patient is asymptomatic, minimally symptomatic or has stable symptoms that can be managed effectively as per the stable CAD treatment guidelines. But physical disruption leads to the formation of a hard thrombus in the vessel

Table 2.1 List of cytokines, chemokine and lipid mediators based on their role in plaque progression or plaque regression

Proinflammatory mediators			Anti-inflammatory mediators		
Cytokines	Chemokines	Lipid mediators	Cytokines	Chemokines	Lipid mediators
IL-1β	MCP-1	LTB$_4$	IL-10	CXCL5	LXA$_4$, ATL
IL-6	CCL3, CCL5	LTC$_4$, LTD$_4$, LTE$_4$	TNF-β	CXCL12	RvD1, RvE1
TNF-α	MIF-1	TXA$_2$		CXCL16	MaR1
IFN-γ	CXCL4, CXCL10			CXCL19	PD1

ATL aspirin-triggered lipoxin, *CCL* C-C chemokine ligand, *CXCL* C-X-C chemokine ligand, *IL* interleukin, *LT* leukotriene, *LX* lipoxin, *MaR1* maresin 1, *MCP-1* macrophage chemoattractant protein 1, *MIF-1* migration inhibitory factor, *PD1* protectin 1, *Rv* resolvin, *TGF* transforming growth factor, *TNF* tumour necrosis factor, *TXA$_2$* thromboxane A2

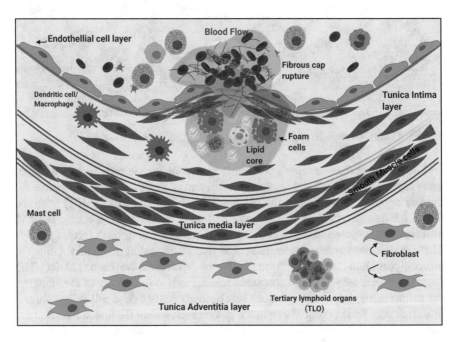

Fig. 2.3 Fracture of the plaque's fibrous capsule, which causes the plaque to be exposed to the blood flow and the blood coagulation components. This event is called thrombosis, the ultimate complication of atherosclerosis which can cause obstruction of the blood flow [13, 26, 27]. (Created with BioRender.com)

lumen that can impede blood flow and rapidly transition to an acute coronary syndrome (ACS). The progression of atherosclerosis from stable CAD through to acute coronary syndrome and ultimately cardiac death if not treated is depicted in Fig. 2.4 [28]. All the risk factors explained earlier act either directly or indirectly at several points along this atherosclerotic pathogenic pathway.

In addition to lipid accumulation and inflammation, vascular calcification also plays an important role in atherosclerosis. Coronary calcification is an active

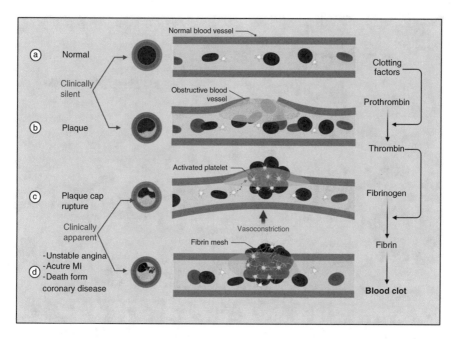

Fig. 2.4 Progression of atherosclerosis and clinical findings at various stages [29–32]. (Created with BioRender.com)

process in the vessel wall that mimics bone formation and is controlled by complex enzymatic and cellular pathways. It can be quantified easily using a computed tomography (CT) technique such as a coronary artery calcium score (CAC) [29]. Such arterial calcification tends to increase the stiffness of the vascular wall and can be measured as an increased arterial pulse wave velocity (PWV). Microcalcification can occur in the necrotic core when the severe inflammation and necrosis begin to stabilize. The role of calcification on plaque stability or instability is still a topic of debate in the medical community, but it is known to involve active reprogramming of vascular smooth muscle cells (VSMCs) by local environmental cues into a dynamic range of phenotypes [33–36]. A proposed schematic diagram showing inflammatory and calcification activity within atherosclerotic lesions with ^{18}F-fluoro-deoxyglucose (FDG) and ^{18}F-sodium fluoride (NaF) as imaging biomarkers is depicted in Fig. 2.5 [33, 37, 38].

When the plaque ruptures and thrombosis occurs, this leads to a lack of blood flow and oxygen supply to the heart, which causes the cessation of aerobic metabolism, depletion of creatine phosphate and the onset of anaerobic glycolysis. This sequence of adverse events starts the ischemic cascade, which is followed by the accumulation of tissue lactate, a progressive reduction in tissue adenosine triphosphate (ATP) levels and an accumulation of catabolites, including those of the adenine nucleotide pool. As ischemia continues, tissue acidosis develops and an efflux of potassium occurs in the extracellular space. In addition, ATP levels fall below those required to maintain critical membrane function, resulting in the onset of heart muscle myocyte death. This may lead to either reversible or irreversible tissue injury

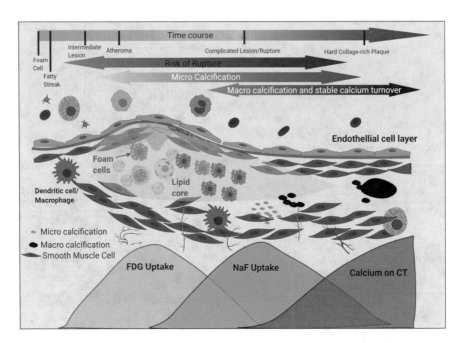

Fig. 2.5 Time course of inflammation and calcification in atherosclerosis. Inflammation and calcification activity can be seen within atherosclerotic lesions with ^{18}F-fluoro-deoxyglucose (FDG) and ^{18}F-sodium fluoride (NaF) as imaging biomarkers. Inflammation is the predominant mechanism active within plaque during early stages of atherosclerosis and FDG may be taken up by the lesion. As inflammation peaks, the risk of plaque rupture may increase. Inflammation also initiates calcium metabolism within lesions resulting in the formation of early calcium deposits that can be seen by uptake of both FDG and hydroxyapatite-specific ^{18}F-sodium fluoride (NaF). It becomes visible with CT once the density of calcium deposits exceeds a certain threshold. During active calcification, plaque may still be vulnerable and eventually, the calcification and mineralization processes exceed the inflammatory activity present within plaque, which might be demarcated by only NaF uptake (in the absence of FDG), as well as calcium deposits on CT. Ongoing calcification eventually leads to forming an end-stage stable atheroma that is densely calcified with only evidence for calcium on CT [33, 37, 38]. (Created with BioRender.com)

that depends on its transmural location, residual coronary flow, and the hemodynamic determinants of oxygen consumption. Irreversible myocardial injury begins after 20 min of coronary occlusion if there is a limited collateral supply. It starts at the subendocardium layers and continues to the subepicardial layers. Irreversible cardiac injury can take from 20 min to up to 6 h to occur depending on collateral blood flow as well as other clinical, preconditioning and pathological factors [5, 6, 39].

Acute coronary syndrome (ACS) develops in the form of unstable angina or a myocardial infarction (MI), which can be either non-ST-elevated myocardial infarction (NSTEMI) or ST-elevated myocardial infarction (STEMI) as described in Fig. 2.6. Both types of MI need to be treated as soon as possible [40, 41]. Diagnostic tools, such as blood biomarkers and electrocardiography (ECG), which are explained in detail later in this chapter, are used to determine the type of ACS. An ECG is a recording of the electrical activity of the heart. It is a simple, noninvasive procedure whereby electrodes are placed on the skin of the chest and connected to an instrument that measures electrical activity in and around the heart [41].

Unstable angina is diagnosed when the patient presents with severe chest pain, shows a specific ECG and the blood test does not show markers for a heart attack. Such unstable angina is the precursor to a heart attack and should be treated as soon as possible. As described above, MI can be one of two types: STEMI or NON-STEMI. The STEMI is considered more severe and life threatening, and so needs to be treated before it can cause irreversible damage to the heart muscle. Both STEMI and NSTEMI patients have the same symptoms of unstable angina with elevated biomarkers, but the ECG is abnormal and is different for each condition as shown in Fig. 2.7. This figure shows how lethal the syndrome is and how it affects the cardiac muscle conduction system [6, 43–46].

There are three waveforms generated during each heartbeat by the electrical conduction system of the heart, which can be recorded in the form of an ECG graph as shown in Fig. 2.7 as this graph consists of a P, QRS complex and a T wave and is observed as P, Q, R, S and T wave components. In a normal ECG, the P wave is the first short upward movement followed by the Q wave with a downward deflection, the R wave with a peak, the S wave with a downward falling wave, and the T wave with a modest upward movement. The P wave is associated with right and left atrial depolarization. The QRS wave represents ventricular depolarization and

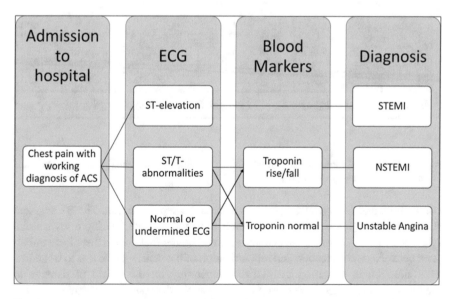

Fig. 2.6 Acute coronary syndrome (ACS) can be classified as unstable angina, STEMI and NSTEMI as determined by various diagnostic tools such as electrocardiography (ECG) and blood biomarkers. *ECG* electrocardiography, *STEMI* ST-elevated myocardial infarction, *NSTEMI* non-ST-elevated myocardial infarction [40–42]

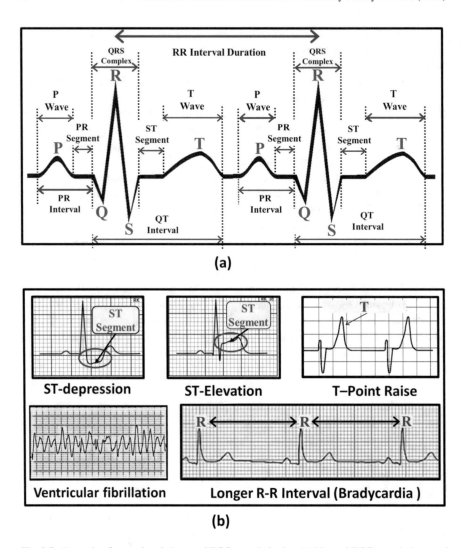

Fig. 2.7 Example of normal and abnormal ECG morphologies. (**a**) Normal ECG morphology and (**b**) abnormal ECG morphology [46]

contraction, and the T wave represents ventricular repolarization. The ST segment starts after the QRS complex and extends until the T wave. During a myocardial infarction this ST segment can become elevated on the ECG when there is an imbalance between depolarization and repolarization of the heart muscle due to ischemia. This situation is classified as STEMI. Depending on the extent of damage to the heart muscle, other wave disruption, such as ST depression or T wave inversion occurs. This is classified as NSTEMI. The magnitude of the ST wave change is associated with the level of risk of the MI.

Chapter 3
Coronary Microvascular Dysfunction (CMD)

Keywords Coronary microcirculation system · Coronary microvascular dysfunction · Coronary blood flow · Angina · Myocardial ischemia

Coronary microvascular dysfunction (CMD) has emerged as an important mechanism of myocardial ischemia in the past two decades. CMD can result from functional and/or structural alterations to the vessel and it results in varying degrees of disruption to the normal coronary physiology as described in Table 3.1.

The intramyocardial arterioles with diameters below 500 μm are actively involved in myocardial perfusion, which is responsible for coronary microcirculation. The epicardial arteries and major arteries are only one segment of the arterial coronary circulation. They are connected to smaller arteries and arterioles that feed the capillaries and cumulatively they are referred to as the coronary microcirculation. This coronary microcirculation system is the main site that regulates the myocardial blood flow, as shown in Fig. 3.1 [47].

The coronary circulation provides the oxygen required by the cardiac pump in order to generate enough adenosine triphosphate (ATP) necessary for systolic contraction. The coronary arterial system, as shown in Fig. 3.1, is composed of three compartments fused together, namely, the proximal compartment of the large epicardial coronary arteries or conductance vessels with diameters ranging from 500 μm up to 2–5 mm, the intermediate compartment of prearteriolar vessels or small arteries with diameters ranging from 100 to 500 μm, and the distal compartment of the arterioles with diameters less than 100 μm. The large epicardial coronary arteries have thick walls and three well-defined layers, as shown in Fig. 2.1. They provide the least resistance to coronary blood flow (CBF) and are further classified into three types: Type I, II and III, based on the number of initial branches

Table 3.1 Pathogenic mechanisms of coronary microvascular dysfunction (CMD) [56]

Alteration	Causes
Structural	
Luminal obstruction	Microembolization in acute coronary syndrome (ACS) or after recanalization
Vascular-wall infiltration	Infiltrative heart disease (e.g. Anderson–Fabry cardiomyopathy)
Vascular remodelling	Hypertrophic cardiomyopathy (HCM), arterial hypertension
Vascular rarefaction	Aortic stenosis, arterial hypertension
Perivascular fibrosis	Aortic stenosis, arterial hypertension
Functional	
Endothelial dysfunction	Smoking, hyperlipidaemia, diabetes
Dysfunction of smooth muscle cell	HCM, arterial hypertension
Autonomic dysfunction	Coronary recanalization
Extravascular	
Extramural compression	Aortic stenosis, HCM, arterial hypertension
Reduction in diastolic perfusion time	Aortic stenosis

and the area of tissue to be supplied with blood as seen in Fig. 3.1. Small arteries and large arterioles are more responsive to flow-dependent dilatation and to changes in intravascular pressure respectively. Large arterioles are responsible for the auto-regulation of the coronary blood flow (CBF) [47, 52–54].

Coronary microvascular dysfunction (CMD) can be caused by various pathogenic conditions and mechanisms, so they are classified clinically as CMD in the presence or absence of myocardial disease, or obstructive CAD, or iatrogenic CMD. This last example could be due to percutaneous coronary intervention (PCI) caused primarily by vasoconstriction or distal embolization. CMD can lead either to impaired dilatation or increased vasoconstriction of the coronary microvessels as a result of these pathogenic conditions [40]. The difference in myocardial ischemia caused by coronary artery stenosis and coronary microvascular dysfunction is depicted in Fig. 3.2 [55]. The pathogenic mechanisms of CMD are described in Table 3.1 [56].

In the case of an epicardial stenosis, the ischemia involves the myocardial territory distal to the stenotic vessel and is more severe in the subendocardium. This is the red area resulting in impairment of contractile function over an extensive zone. In the case of microvascular dysfunction, the ischemia is localized in small myocardial areas that are distributed throughout the myocardial wall (small circles). This does not usually result in detectable impairment of contractile function due to the presence of normal contractile myocardial cells in the same zone [55].

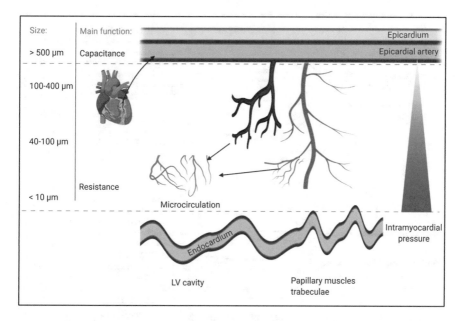

Fig. 3.1 Coronary arterial system. (Diameter: Epicardial arteries >500 μm, main stimulus for vasomotion – flow, Main function – transport; small arteries <500–150 μm, main stimulus for vasomotion – pressure, Main function – regulation; large arterioles <150–100 μm, Arterioles <100 μm, main stimulus for vasomotion – metabolites, Main function – regulation; capillaries <10 μm, exchange as a main function; *LV* Left ventricular) [47–51]. (Created with BioRender.com)

Understanding the cause of CMD is important to understand the abnormal microvascular constriction in patients with normal coronary arteries but who present with chest pain or with symptoms of chronic stable angina. This type of microvascular constriction is an important pathogenetic component of microvascular obstruction (MVO) observed in a high fraction of patients who undergo primary percutaneous coronary intervention (PCI) due to ST-elevated myocardial infarction (STEMI) [57].

Patients who undergo PCI have a high probability for intravascular plugging caused by atherosclerotic debris, micro-emboli and thrombus material typically released during PCI. It also explains why there are ≈20–50% of patients with a prevalence for angina despite successful revascularization surgery [58]. The current goal of PCI or any other revascularization method is to relieve the symptoms rather than improve the pathology, which will require new research in this area if an improvement in CMD pathology is to be achieved [59–62].

In conclusion, among patients with stable or unstable angina, both the symptoms and the myocardial ischemia are caused by a combination of epicardial artery stenosis and CMD. However, the contribution of these two conditions will vary depending on the clinical status of each patient.

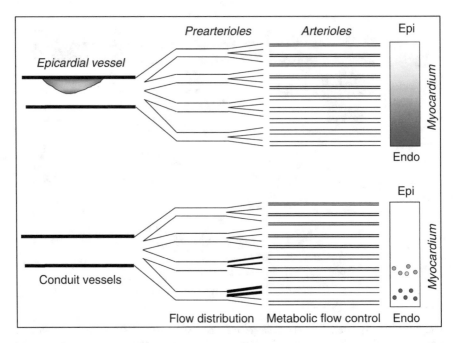

Fig. 3.2 Differences in myocardial ischemia caused by a coronary artery stenosis (upper drawing) or coronary microvascular abnormalities (bottom drawing) [55]

Chapter 4
Diagnosis of Coronary Artery Disease (CAD)

Keywords Diagnosis · Type of angina · Probability of coronary artery disease · Noninvasive testing · Blood-borne biomarkers

Evaluation of the patient with known or suspected cardiovascular disease begins with their medical history and a targeted physical examination as well as basic ancillary studies that are sufficient for the physician to understand the aetiology of any chest pain. A history and symptoms of angina are important in order to determine which tools should be used for diagnosis and treatment. The major signs and symptoms associated with cardiac disease include chest discomfort, dyspnoea, fatigue, oedema, palpitations and syncope. Then the type of angina needs to be characterized based on the clinical classification listed in Table 4.1. Anginal pain is characterized depending on the pain location, the quality, the duration of pain and any exacerbating or alleviating factors [63–65].

If the angina lasts longer and has the symptoms mentioned in Table 4.1, it can be classified as typical angina. If only two symptoms are presented, it can be called atypical angina, and with even fewer symptoms it is described as noncardiac chest pain. Once the physician determines that the appropriate symptoms are present, then the probability of CAD is assessed. For example, a history of CAD in the family, the patient's age, gender and symptoms are all important to determine the probability of having coronary artery disease as shown in Table 4.2. In addition, physicians may use other risk algorithms that are available based on the location and race of the patient.

A description of the symptoms and determining the probability are important elements for the understanding of the occurrence of CAD, its severity, comorbidities and complications. But alone they are not sufficient to diagnose CAD. The Canadian Cardiovascular Society (CCS) has proposed that the appropriate diagnosis and management of stable ischemic heart disease (SIHD) also includes the need for basic ancillary studies, such as fasting lipids, a resting 12-lead electrocardiogram and possibly a chest X-ray. For example, for patients $\gtrsim$ 40 years of age, the

M. W. King et al., *Coronary Artery Disease and The Evolution of Angioplasty Devices*, SpringerBriefs in Materials, https://doi.org/10.1007/978-3-030-42443-5_4

Table 4.1 Clinical classification of chest pain [63]

Type of angina	Symptoms
Typical angina	1. Retrosternal chest discomfort 2. Increased pain with exertion or emotional stress 3. Relief with rest or nitroglycerin
Atypical angina	Exhibits 2 of the above symptoms
Noncardiac chest pain	Exhibit 0 or 1 of the above symptoms

Table 4.2 Probability of coronary artery disease (CAD) by age, gender and symptoms [63]

Age (years)	Nonanginal pain (%)		Atypical angina (%)		Typical angina (%)	
	Women	Men	Women	Men	Women	Men
30–39	5	18	10	29	28	59
40–49	8	25	14	38	37	69
50–59	12	34	20	49	47	77
60–69	17	44	28	59	58	84
70–79	24	54	37	69	68	89
>80	32	65	47	78	76	93

CCS has suggested the use of noninvasive testing in patients with classical anginal chest pain symptoms to diagnose SIHD as explained in Fig. 4.1 and acute coronary syndrome (ACS) as presented in Fig. 2.6 [64]. Depending on the situation, some invasive diagnostic testing can be used by physicians, such as fractional flow reserve (FFR) and intravascular ultrasound, as suggested by the American College of Cardiology Foundation (ACCF) and the American Heart Association (AHA) in their 2012 document describing the guidelines for the appropriate criteria for using diagnostic catheterization [66].

There are many advanced diagnostic tools available now such as stress electro-cardiography, echocardiography, myocardial perfusion imaging, magnetic resonance imaging, coronary computed tomography and cardiac catheterization. The selection of the initial test depends on the patient's characteristics, potential contra-indications to testing, limitations of each modality, local availability and local expertise. Figure 4.2 describes the comprehensive flow chart for the diagnosis as well as treatment of the patient with chest pain. One should take a note that only around 1–11% of patients admitted to hospital with chest pain are due to coronary artery disease (CAD) or acute coronary syndrome (ACS). Physicians might also use the sensitivity and specificity data of any diagnostic test, as listed in Table 4.3, to decide which test is to be used for which patient [64, 67].

For a suspected CAD patient, a treadmill exercise test with a 12-lead ECG and a blood pressure monitor is a useful option as it is simple, low cost and easily available. But pharmacological testing with vasodilator perfusion imaging or dobutamine echocardiography is preferred for those who cannot exercise to an adequate workload due to various reasons such as obesity, orthopaedic limitations, balance

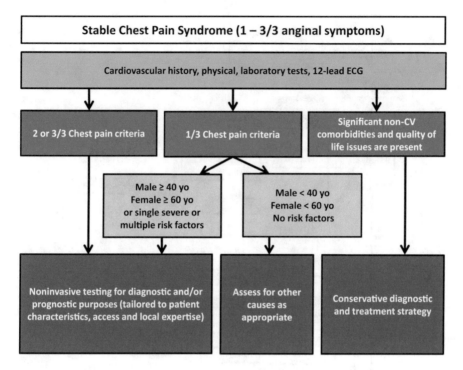

Fig. 4.1 Noninvasive testing for diagnosis and prognostic purposes in patients with classical chest pain symptoms (*CV* cardiovascular, *ECG* electrocardiogram) [64]

issues, pulmonary limitations, frailty or limb dysfunction. Vasodilator perfusion imaging or anatomical imaging for diagnostic purposes is an appropriate method when complete left bundle branch block (LBBB) or paced ventricular rhythm is present. Computed tomography (CT) can be used to detect coronary calcium or to generate a coronary angiogram, but coronary computed tomography angiography (CCTA) is preferable for individuals who have a probability of being in an intermediate risk category for CAD. One thing to consider is that CCTA should also be avoided for patients with arrhythmia, significant renal dysfunction or contrast media allergies. Invasive coronary angiography is the benchmark investigative technique to detect the presence of CAD causing luminal blockage, but not for the detection of early atheroma. It should not be offered to patients who are not candidates for revascularization [64, 68–70].

Other than the above-mentioned diagnostic tools, blood-borne biomarkers are also helpful to access the diagnosis and prognosis as well as to monitor the successful treatment of CAD. For example, inflammatory biomarkers such as high sensitivity C-reactive protein (CRP) and interleukin-6 are tools for the prognosis of future cardiac events. Biomarkers of myocardial injury, such as troponin T and I levels,

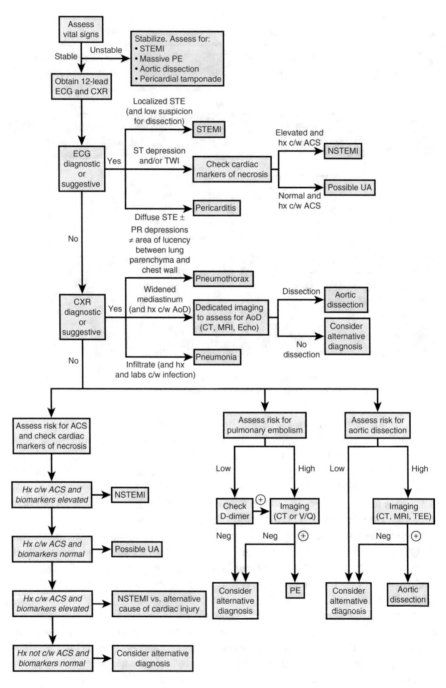

Fig. 4.2 A comprehensive approach to the diagnosis and treatment of patients with chest pain

Table 4.3 Summary estimates of pooled sensitivity and specificity data for cardiac tests (with 95% confidence interval) used in the diagnosis of CAD [64]

Technology[a]	Sensitivity	Specificity
Exercise treadmill	0.68 (0.23–1.0)	0.77 (0.17–1.0)
Attenuation-corrected SPECT	0.86 (0.81–0.91)	0.82 (0.75–0.89)
Gated SPECT	0.84 (0.79–0.88)	0.78 (0.71–0.85)
Traditional SPECT	0.86 (0.84–0.88)	0.71 (0.67–0.76)
Contrast stress echocardiography (wall motion)	0.84 (0.79–0.90)	0.80 (0.73–0.87)
Exercise or pharmacologic stress echocardiography	0.79 (0.77–0.82)	0.84 (0.82–0.86)
Cardiac computed tomographic angiography	0.96 (0.94–0.98)	0.82 (0.73–0.90)
Positron emission tomography (PET)	0.90 (0.88–0.92)	0.88 (0.85–0.91)
Cardiac MRI (perfusion)	0.91 (0.88–0.94)	0.81 (0.75–0.87)

[a]*MRI* magnetic resonance imaging, *SPECT* single photon emission computed tomography

also have potential prognosis features associated with stable CAD, and are specific markers for myocardial injury. Biomarkers of vascular function and neurohumoral activity, such as B-type natriuretic peptide (BNP) and the N-terminal fragment of its prohormone named NT-proBNP, can be used for risk assessment related to vasoactive function, which can be an important surrogate for determining the severity of heart failure. Biomarkers such as Atrial Natriuretic Peptide (ANP), Adrenomedullin (ADM) and Growth Differentiation Factor-15 (GDF-15), novel omics-based biomarkers of renal function such as Estimated Glomerular Filtration Rate (eGFR) and Cystatin C, and lipid biomarkers such as total cholesterol (TC), high density lipoprotein cholesterol (HDL-C), LDL-C and triglycerides (TGs), can all play a part in determining the risk level of suspected CAD patients as illustrated in Fig. 4.3 [71].

All the above-mentioned biomarkers have their own specific roles and characteristics for the diagnosis and prognosis of CAD but they are not yet recommended for routine clinical practice due to their limitations compared to other established diagnostic tools. Advances in research continue to make them more effective, which will enable them to be translated into clinical practice in the future.

Fig. 4.2 (continued) *ACS* acute coronary syndrome, *CT* computed tomography, *CXR* chest X-ray, *ECG* electrocardiogram, *MRI* magnetic resonance imaging, *NSTEMI* non-ST-elevated myocardial infarction, *PE* pulmonary embolism, *ST* is a segment of the ECG pulse and is not defined, *STE* ST elevation, *STEMI* ST-elevated myocardial infarction, *TEE* transoesophageal echocardiogram, *TWI* T wave inversion, *UA* unstable angina, *V/Q* ventilation/perfusion, *Hx* history, *c/w* consistent with [67]

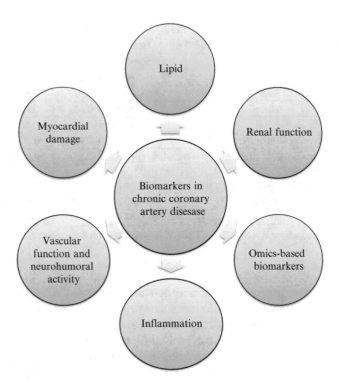

Fig. 4.3 Established and novel biomarkers for chronic coronary artery disease [71]

Chapter 5
Treatment of Coronary Artery Disease (CAD)

Keywords Treatment · Drug therapy · Invasive therapy · Minimally invasive therapy · Therapeutic guideline · Dual antiplatelet therapy · Coronary artery bypass grafting

The treatment of coronary artery disease (CAD) aims to treat cardiac disease based on its severity and pathological condition. Along with pain, patients may suffer from other symptoms such as severe fatigue, dyspnoea, abdominal pain, nausea and sweating, and to understand cardiac pain one requires knowledge of the interplay between ischemic, metabolic and neurological mechanisms behind the CAD. This chapter will discuss the therapies for stable ischemic heart disease (SIHD) and for acute coronary syndrome (ACS).

For the patient with SIHD, there are five fundamental aspects that need to be followed alone and/or in combination with each other:

1. Identification, treatment and education of the patient
2. Reduction of coronary risk factors
3. Secondary prevention using pharmacologic and/or non-pharmacologic intervention, with attention to adjustments in lifestyle
4. Pharmacologic management of angina
5. Revascularization by catheter-based percutaneous coronary intervention (PCI) or by coronary artery bypass grafting (CABG)

Identification of the cause of SIHD is necessary in order to apply the right therapy. For example, there are several medical conditions which either lead to an increase in myocardial oxygen demand or a reduction in oxygen delivery. Both these conditions contribute to the onset of new angina pectoris as shown in Fig. 5.1 [39].

Reduction of coronary risk factors can also help to manage SIHD effectively. For example, the management of hypertension can reduce CAD events and mortality by up to 16%. Other factors such as cigarette smoking, management of dyslipidaemia, high-density lipoprotein and cholesterol levels, obesity and inflammation have

M. W. King et al., *Coronary Artery Disease and The Evolution of Angioplasty Devices*, SpringerBriefs in Materials, https://doi.org/10.1007/978-3-030-42443-5_5

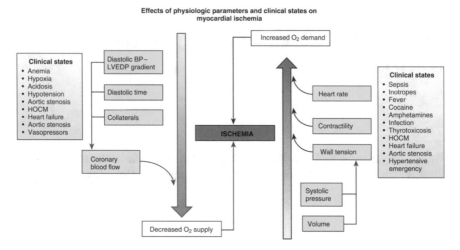

Fig. 5.1 Participants with myocardial ischemia. The hemodynamic consequences of clinical states and their effects on the supply and demand of oxygen and ultimately on ischemia [39] *BP* blood pressure, *HOCM* hypertrophic obstructive cardiomyopathy, *LVEDP* left ventricular end-diastolic pressure

also been shown to reduce CAD events and mortality [6, 28]. Antianginal therapy helps to reduce symptoms and helps to prolong the ability to exercise. Therapies such as aspirin, angiotensin-converting enzyme (ACE) inhibitors and lipid-lowering treatments have been shown to reduce mortality and morbidity in patients with SIHD. Other therapies such as long-acting nitrates, beta blockers or calcium channel blockers improve the symptoms and exercise performance, but do not have a beneficial effect on improving the survival of patients. The antianginal drug therapies are listed in Table 5.1 [28].

Beta blockers have anti-ischemic, antiarrhythmic and antihypertensive properties. Beta blockade reduces myocardial O_2 requirements by slowing the heart rate and helping to reduce the exercise-induced blood pressure. Calcium antagonists inhibit calcium ion movement through the slow channels in the cardiac and smooth muscle membranes by noncompetitive blockade of voltage-sensitive L-type calcium channels that lead to a reduction in myocardial O_2 demand and an increase in O_2 supply. Nitrates relax vascular smooth muscle. Nitrates exhibit a vasodilatory effect, which reduces the ventricular preload, leading to reduced myocardial wall tension and O_2 requirements. This makes them a useful drug for heart failure and angina. Other than these conventional drugs some novel antianginal drugs, such as ranolazine, ivabradine, nicorandil, trimetazidine and molsidomine, also have anti-ischemic properties. Ranolazine reduces myocardial ischemia through a reduction in calcium overload in ischemic myocytes by inhibiting the inward movement of sodium. Ivabradine is a heart rate-slowing drug, nicorandil increases nitric oxide concentration that leads directly to coronary vasodilation. Trimetazidine increases myocardial glucose utilization and prevents adenosine triphosphate (ATP) reduction.

Table 5.1 Antianginal drugs based on their effect with side effects and contraindications [28]

Agents	Side effects	Contraindications
Agents with a physiologic effect		
Short- and long-acting nitrates	Headache, flushing, hypotension, syncope and postural hypotension, reflex tachycardia, methaemoglobinaemia	Hypertrophic obstructive cardiomyopathy
Beta blockers	Fatigue, depression, bradycardia, heart block, bronchospasm, peripheral vasoconstriction, postural hypotension, impotence, masked signs of hypoglycemia	Low heart rate or heart conduction disorder, cardiogenic shock, asthma, severe peripheral vascular disease, decompensated heart failure, vasospastic angina; use with caution in patients with COPD (cardioselective beta blockers may be used if the patient receives adequate treatment with inhaled glucocorticoids and long-acting beta agonists)
Calcium channel blockers		
Nondihydropyridine (heart rate–lowering agents)	Bradycardia, heart conduction defect, low EF, constipation, gingival hyperplasia	Cardiogenic shock, severe aortic stenosis, obstructive cardiomyopathy
Dihydropyridine	Headache, ankle swelling, fatigue, flushing, reflex tachycardia	Low heart rate or heart rhythm disorder, sick sinus syndrome, CHF, low blood pressure
Ivabradine	Visual disturbances, headache, dizziness, bradycardia, atrial fibrillation, AV block	Low heart rate or heart rhythm disorder, severe hepatic disease; not to be prescribed with verapamil and diltiazem; caution for use in patients with AF
Nicorandil	Headache, facial flushing, dizziness and weakness, nausea, hypotension; oral, anal or gastrointestinal ulceration	Cardiogenic shock, heart failure, low blood pressure (<100 mm Hg systolic)
Molsidomine	Headache, hypotension	None reported
Agents that affect myocardial metabolism		
Ranolazine	Dizziness, constipation, nausea, QT-interval prolongation	Liver cirrhosis, long QT interval on ECG test
Trimetazidine	Gastric discomfort, nausea, headache, movement disorders	Allergy, Parkinson disease, tremors, movement disorders, severe renal impairment
Perhexiline	Dizziness, nausea, vomiting, lethargy, tremors	Slow hydroxylators of cytochrome P450, abnormal liver function, neuropathy
Allopurinol	Rash, gastric discomfort	Hypersensitivity, renal failure

AF atrial fibrillation, *AV* atrioventricular, *CCB* calcium channel blocker, *CHF* congestive heart failure, *COPD* chronic obstructive pulmonary disease, *ECG* electrocardiography, *EF* ejection fraction

Molsidomine reduces preload, dilates the coronary arteries and increases the donation of nitric oxide. All the medical treatments described above should be used alone or in combination with others based on the patient's need and careful assessment. To achieve this, the physician may use a guideline, such as the National Institute for Health and Care Excellence (NICE) guideline published by the National Clinical Guideline Centre, for the management of stable angina as illustrated in Fig. 5.2 [72].

In the past decade, treatments using living cell-based therapy and gene therapy research are booming due to advanced technologies and increased understanding of the vascular and cellular architecture at the molecular level. There are currently many published research studies using vascular endothelial growth factor (VEGF), fibroblast growth factor (FGF) as well as pluripotent stem cells that have proven their safety, angiogenesis and advanced regenerative capacity. At the same time, there is concern that the direct or indirect use of living cells to produce growth factors on the device's surface will limit their long-term therapeutic effect and cause side effects specifically host cell rejection. In order to overcome this limitation, Dr. Ke Cheng and his research group at BioTherapeutics Laboratory at North Carolina State University have demonstrated the use of novel stem cell-like micro and nano particles (CMMP and CMNP) to initiate angiogenesis and the therapeutic regenerative properties of stem cells without using any actual living cells [73–75].

The principle guideline for the management of all patients has two main goals: medical management and risk factor reduction. The procedure for revascularization, such as coronary artery bypass grafting (CABG) and percutaneous coronary intervention (PCI) in patients with stable ischemic heart disease, is a topic of debate over the last two decades. On the one hand, PCI and CABG are the most frequently applied treatments for STEMI patients, but for the stable ischemic heart disease patient there are many more factors involved before revascularization can be considered. Such patients, such as those presenting with underlying risk factors, sociodemographic factors like age, and physical capacity, have the ability to adhere to

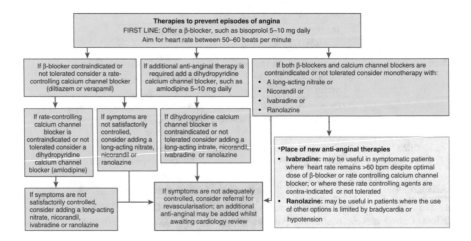

Fig. 5.2 2011 NICE guidelines recommending medical therapy for chronic stable angina [72]

prescribed treatments, lifestyle interventions, overall quality of life, other medical conditions and the patient's preferences. The presence and severity of symptoms, the physiologic conditions of coronary lesions and other anatomical considerations, myocardial ischemia and the presence of LV dysfunction, along with other medical conditions are major deciding factors for the selection of revascularization of patients with stable ischemic heart disease. Such revascularization is normally considered after intensive medical therapy and/or risk factor modification and other conditions that limit the extent of antianginal therapy [28, 76, 77].

For both the NSTEMI and STEMI patient, the goal of treatment is the immediate relief of ischemia and the prevention of MI and death. The patient is treated immediately with antianginal, antiplatelet, fibrinolytic and anticoagulant therapies, and the patient with severe continuing angina or a large MI and other LV functions is managed by CABG or PCI along with continuous medical therapy. Irrespective of the indication for revascularization, PCI should be coupled with optimal medical therapy after the procedure, such as the control of risk factors and other medical therapies as per the published guidelines. Control of hypertension and diabetes, exercise and smoking cessation, lipid management and statin therapy are all important components of optimal medical therapy. CABG has the advantage of a late mortality benefit compared to PCI, but early procedural risks and longer in-hospital recovery periods are the major factors to be considered while selecting PCI or CABG along with other critical factors [78, 79].

Medical therapies involve supplemental oxygen, nitrates, analgesic therapy, beta-adrenergic blockers, calcium channel blockers, the management of cholesterol, inhibitors of the renin-angiotensin-aldosterone system, antiplatelet and anticoagulant therapies alone or in combination. These options are described in the published guidelines and are followed by revascularization if required, as shown in Figs. 5.4 and 5.5 and the 2014 American Heart Association/American College of Cardiology (AHA/ACC) guidelines for the management of patients with NSTEMI (also called NSTE-ACS) and the 2013 guideline for management of patients with STEMI [42, 80]. To prevent such chronic CAD, a wide range of antithrombotic, single and dual antiplatelet therapies are prescribed and the mechanism of actions (MOA) is depicted in Fig. 5.3 [81].

The activation of platelets occurs when the first blood agents adhere to ruptured or eroded surfaces of the plaque, aggregate at the site and then ultimately start the coagulation cascade. Drugs like aspirin, as shown in Fig. 5.3, block the cyclooxygenase-1 enzyme which promotes the synthesis of thromboxane A_2 receptor expressed in platelets and other inflammatory cells, and this is the first line of treatment for suspected ACS patient. $P2Y_{12}$ inhibitors such as thienopyridines (ticlopidine, clopidogrel and prasugrel), ticagrelor (cyclopentyl-triazolo-pyrimidine CPTP inhibitor) and cangrelor (ADP inhibitor) prevent platelet aggregation by inhibiting the release of the platelet adenosine diphosphate (ADP) receptor $P2Y_{12}$. Dual antiplatelet therapy involving a combination of aspirin and $P2Y_{12}$ receptor blocker such as clopidogrel has become the standard procedure of care for patients undergoing revascularization. GP IIb/IIIa antagonists, such as Abciximab, Eptifibatide, Tirofiban, interfere with platelet cross-linking and platelet-derived thrombus formation [81–84].

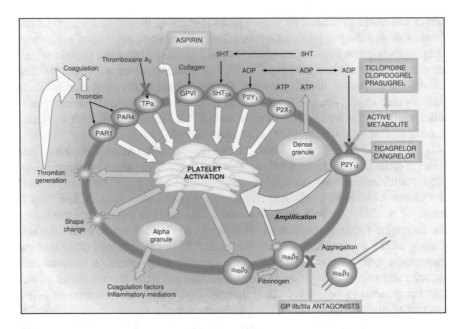

Fig. 5.3 Platelet activation and aggregation inhibitors [81] *ADP* adenosine diphosphate, *ATP* adenosine triphosphate, *GP* glycoprotein, *HT* hydroxytryptamine, *PAR* protease activated receptor; *TP* thromboxane A$_2$ receptor

As described in Figs. 5.4 and 5.5, the timeline to undergo revascularization for the patient with ACS is important. Early PCI helps the coronary system to restore blood circulation and limits any permanent damage to the heart muscle or, in other words, irreversible myocardial injury. As shown in Fig. 5.5, and as per the guideline, once admitted to a PCI-capable hospital, the first medical contact (FMC) to PCI time should be less than 90 min for STEMI patients, and if admitted to a non-PCI-capable hospital and fibrinolytic agents have not been administered, the FMC to PCI time should be less than 120 min.

Such a revascularization technique is the therapy of choice for ST-elevated myocardial infarction (STEMI) patients. Approximately 95% of patients are treated with PCI to get their blood flow restored by opening a blocked artery, compared with

Fig. 5.4 (continued) See corresponding full-sentence recommendations and their explanatory footnotes in 2014 AHA/ACC guideline for the management of patients with non-ST-elevated acute coronary syndrome (NSTE ACS). †In patients who have been treated with fondaparinux (as upfront therapy) who are undergoing percutaneous coronary intervention (PCI), an additional anticoagulant with anti-IIa activity should be administered at the time of PCI because of the risk of catheter thrombosis. *ASA* indicates aspirin, *CABG* coronary artery bypass graft, *cath* catheter, *COR* class of recommendation, *DAPT* dual antiplatelet therapy, *GPI* glycoprotein IIb/IIIa inhibitor, *LOE* level of evidence, *NSTE-ACS* non-ST-elevated acute coronary syndrome, *PCI* percutaneous coronary intervention, *pts.* patients, *UFH* unfractionated heparin

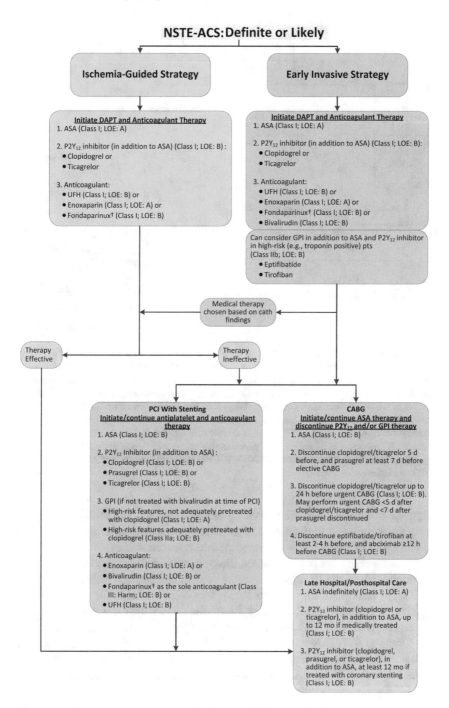

Fig. 5.4 Algorithm for management of patients with definite or likely non-ST-elevated acute coronary syndrome (NSTE-ACS) [42]

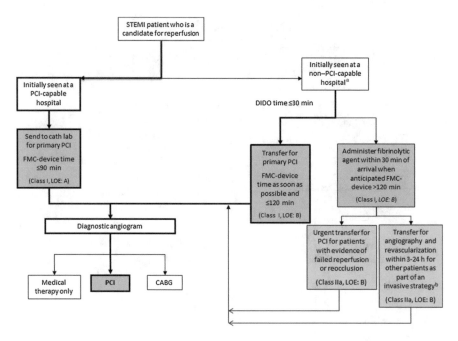

Fig. 5.5 Reperfusion therapy for patients with STEMI [80] [a]The bold arrows and boxes are the preferred strategies. Performance of PCI is dictated by an anatomically appropriate stenosis. Patients with cardiogenic shock or severe heart failure initially seen at a non-PCI-capable hospital should be transferred for cardiac catheterization and revascularization as soon as possible, irrespective of the delay since the onset of MI (Class I, LOE: B). [b]Angiography and revascularization should not be performed within the first 2–3 h after administration of fibrinolytic therapy. *CABG* coronary artery bypass graft, *DIDO* door-in-door-out, *FMC* first medical contact, *LOE* level of evidence, *MI* myocardial infarction, *PCI* percutaneous coronary intervention, *STEMI* ST-elevated myocardial infarction. Please refer to the 2013 ACCF/AHA guideline for patients with STEMI for LOE and classification details

only 54% of patients who are treated by medical therapy. The evolution in angioplasty techniques is described later in the next chapter together with a detailed discussion about PCI techniques and new updated research studies.

As described earlier in this chapter, there are certain risk-benefit ratios involved in recommending the therapies regarding revascularization, whether it be PCI or CABG. For example, Table 5.2 lists the appropriate use criteria for PCI and CABG in patients with multivessel coronary artery disease, whether it is A – Appropriate, I – Inappropriate or U – Uncertain to apply the particular revascularization procedure using the SYNTAX score. The SYNTAX score, by combining anatomical and clinical prognostic variables, creates accurate mortality predictions to guide the choice between PCI and CABG for patients with multivessel coronary disease [85].

Table 5.2 Appropriate use criteria for common indications in patients with multivessel coronary disease [85]

Multivessel coronary disease	CABG	PCI
Two-vessel CAD with proximal LAD stenosis	A	A
Three-vessel CAD with low CAD burden (i.e. 3 focal stenosis, low SYNTAX score)	A	A
Three-vessel CAD with intermediate to high CAD burden (i.e. multiple diffuse lesions, presence of CTO or high SYNTAX score)	A	U
Isolated left main stenosis	A	U
Left main stenosis and additional CAD with low CAD burden (i.e. one to two vessel additional involvement, low SYNTAX score)	A	U
Left main stenosis and additional CAD with intermediate to high CAD burden (i.e. three vessel involvement, presence of CTO or high SYNTAX score)	A	I

A appropriate, *I* inappropriate, *U* uncertain, *CABG* coronary artery bypass grafting, *CAD* coronary artery disease, *CTO* chronic total occlusion, *LAD* left anterior descending artery, *PCI* percutaneous coronary intervention, *SYNTAX* synergy between PCI with TAXUS and cardiac surgery

Chapter 6
Evolution of Angioplasty Devices

Keywords Angioplasty device · Balloon catheter · Stent · Drug-eluting stent · Resorbable stent · Cell-based therapy · FDA-approved devices · Commercial devices · Drug polymer coating · Antiproliferative drugs · Experimental devices

It is estimated that over 600,000 to 1 million cardiac catheterizations (CC) are performed annually in the United States. This exceeds the number of coronary artery bypass graft procedures (CABG) which are growing at an annual growth rate of about 1–5% in the United States. Percutaneous coronary intervention (PCI) is the treatment to open the blocked coronary artery using a transcatheter intervention that involves the insertion of a catheter through the femoral or radial artery and then guiding it to the site of stenosis and opening the blocked artery by inflating the angioplasty device such as a balloon catheter, bare metal stent or drug-eluting stent procedure, as classified in Fig. 6.1 [6, 86].

After Dr Mason Sones discovered selective coronary angiography accidentally in 1958, Andreas Gruentzig, a German-born physician in Zurich, Switzerland, performed the first balloon angioplasty procedure using a fixed-wire catheter in a coronary artery in 1977, and eventually the first stent implantation was performed in a patient by Sigwart and colleagues 1 year later [87–89]. Figure 6.2 shows the developments in angioplasty devices over the last 40 years [90].

During the development and evolution of these PCI devices, restenosis and thrombosis are the two major clinical complications that have been observed, and both conditions are due to the type of material used for the implanted scaffold along with other mechanical and biological risk factors. In simple terms, restenosis is a gradual re-narrowing of the stented segment that occurs most often between 3 and 12 months after stent placement. It usually presents as recurrent angina, but it can present as an acute myocardial infarction, which should be managed by repeat percutaneous revascularization. In contrast, stent thrombosis is an abrupt thrombotic occlusion of the vessel because of impaired or delayed healing, and this results in a catastrophic complication that presents either as a large myocardial infarction or as sudden death.

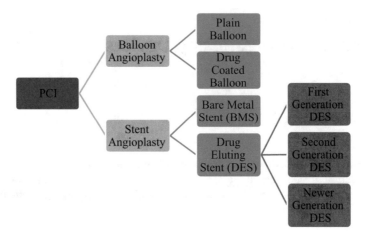

Fig. 6.1 Percutaneous coronary intervention (PCI) angioplasty procedure classification

There are several additional risk factors for late stent thrombosis, such as the penetration of the necrotic core, malapposition, overlapping stent placement, excessive stent length and bifurcated lesions. These factors represent additional barriers to healing, during drug-eluting stent (DES) implantation as mentioned in Fig. 6.3. They should be avoided so as to minimize the risk of thrombosis. Four categories of stent thrombosis have been defined as acute (0–24 h), early or subacute (within 30 days), late (between 30 days and 1 year) and very late (more than 1 year). We will discuss in this chapter how in the last three decades, starting from balloon angioplasty to the latest novel stent platform, there has been progress to diminishing restenosis, thrombosis and other limitations for each generation of device [91–94].

Balloon Catheter System for Angioplasty

Balloon catheters are used for primary percutaneous transluminal angioplasty (PTA) either with or without a stent that is crimped to it and are available in a wide variety of sizes, lengths, shapes and material compositions. The first generation of balloon catheters had a fixed-wire catheter-based balloon which then transitioned to an over-the-wire and exchangeable system (Rapid Exchange) over the last decade, which now allows the guide wire and balloon to move independently. An inflation device with an attachment hub is used at the proximal end of the balloon catheter near the site of insertion to inflate the balloon as shown in Fig. 6.4.

For any balloon catheter device, there are three clinical goals: deliverability, crossability and dilatation. But two major limitations of bare balloon catheter devices led to the development of the next generation of drug-eluting balloons and the first generation of a stent angioplasty scaffolds. The first limitation was the over-the-wire exchangeable system, which led to early closure of the treated vessel within a few hours to days that required repeated dilatation or emergency coronary artery bypass

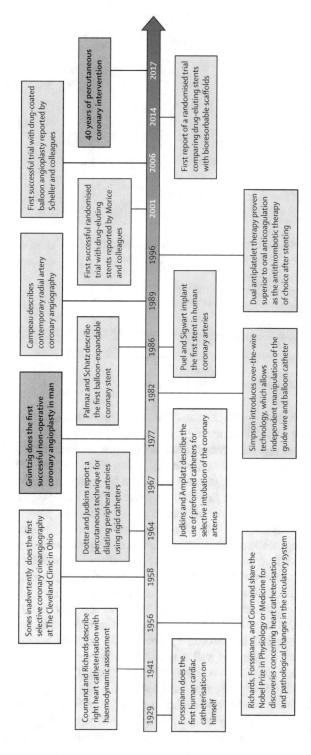

Fig. 6.2 Timeline of diagnostic cardiac catheterization, coronary balloon angioplasty, stent and scaffold implantation as contributors make improvements to reduce cardiovascular deaths. Developments in diagnostic catheterization are shown in green, coronary angioplasty in red and catheter therapeutics in blue [90]

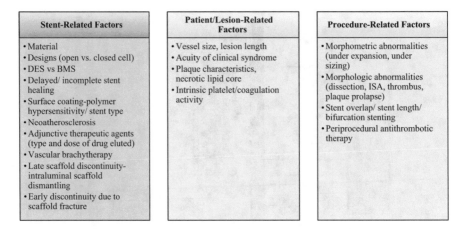

Stent-Related Factors	Patient/Lesion-Related Factors	Procedure-Related Factors
• Material • Designs (open vs. closed cell) • DES vs BMS • Delayed/ incomplete stent healing • Surface coating-polymer hypersensitivity/ stent type • Neoatherosclerosis • Adjunctive therapeutic agents (type and dose of drug eluted) • Vascular brachytherapy • Late scaffold discontinuity-intraluminal scaffold dismantling • Early discontinuity due to scaffold fracture	• Vessel size, lesion length • Acuity of clinical syndrome • Plaque characteristics, necrotic lipid core • Intrinsic platelet/coagulation activity	• Morphometric abnormalities (under expansion, under sizing) • Morphologic abnormalities (dissection, ISA, thrombus, plaque prolapse) • Stent overlap/ stent length/ bifurcation stenting • Periprocedural antithrombotic therapy

Fig. 6.3 Precipitants of stent thrombosis [94] *BMS* bare metal stent, *CHF* congestive heart failure, *DES* drug-eluting stent, *ISA* incomplete stent strut apposition

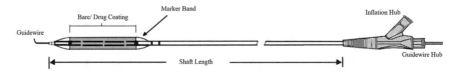

Fig. 6.4 Illustration of over-the-wire (OTW) balloon catheter

grafting (CABG) in about 3–5% of cases. The second limitation was the high rate of restenosis resulting in the recurrence of symptoms in about 20–30% of patients, mainly due to plaque prolapse, vessel recoil and constrictive remodelling. Table 6.1 contains the list of polymers utilized in the fabrication of the balloons [95, 96].

The first balloon used by Dr Gruentzig was made from polyvinyl chloride (PVC) film, but it had a thick wall. Over time balloon materials and technologies evolved to use improved thinner polymer materials such as polyethylene (PE), polyethylene terephthalate (PET) and nylon. PET offers the advantages of tensile strength and maximum pressure rating, while nylon is softer. There have been developments over the years in terms of surface coatings of angioplasty balloons to improve lubrication, trackability and abrasion resistance and to deliver an antirestenotic/anticoagulatory drug. A variety of different balloon coatings that either modify the surface properties or release an active pharmaceutical ingredient (API) have been reported. They include lubricious coatings, both hydrophilic and hydrophobic, abrasion and puncture resistant coatings, tacky or high friction coatings, conductive coatings, antithrombogenic coatings, drug release coatings, as well as reflective and selective coatings [96, 98].

A drug-coated balloon (DCB) or a drug-eluting balloon (DEB) is used to release an active pharmaceutical ingredient at the implantation site during the short duration of contact between the balloon surface and the site of injury without using a permanent metal stent. An early generation of balloon catheter systems utilized drugs such as paclitaxel and sirolimus. Among these drugs, paclitaxel is the

Table 6.1 Basic materials and properties of balloons [95, 97]

Balloon material	Compliance (%)	Burst pressure	Balloon selection based on compliance	Scratch resistance	Max. rated pressure	
					ATM	PSI
Polyvinyl chloride	High (>10)	Moderate	Pre-dilatation	Unknown	6–8	88–117
Polyethylene	Moderate (>10)	Moderate	Pre-dilatation	Low-moderate	10	147
Polyolefin copolymers	High (>10)	Moderate	Pre-dilatation	Unknown	Unknown	Unknown
Polyethylene terephthalate	Non (<5)	Highest	Post-dilatation of stents, resistant lesion	Low	20	294
Nylon	Low (5–10)	High	Pre-dilatation/post-dilatation of stents	Moderate	16	235
Nylon-reinforced polyurethane	Non (<5)	High	Post-dilatation of stents, resistant lesion	High	10	147
Polyurethane	Low (5–10)	High	Pre-dilatation/post-dilatation of stents	Unknown	10	147

preferred API due to its hydrophobicity, rapid drug uptake and retention. The mechanism of action of the drugs used to coat balloon and stent systems is discussed later. DCB's have not been approved for patients with a myocardial infarction; however, the FDA has approved the use of DCB's for peripheral artery disease (PAD) as mentioned in Table 6.2.

Clinically, data from 23 clinical trials involving a total of 2712 patients have compared drug-coated balloons with drug-eluting stents for the treatment of CAD. The data shows that DCB is equivalent to DES in terms of safety for managing CAD. Some other clinical trials have confirmed that the DCB leads to fewer incidents of in-stent restenosis and the occlusion of small coronary vessels [103–107]. A list of some CE-mark approved drug-coated balloon (DCB) devices is mentioned in Table 6.3. The use of DCB's for the treatment of bare metal stents-in-stent restenosis (BMS-ISR) or DES-in-stent restenosis (DES-ISR) has proven to be superior to plain balloon's and the first generation of DES angioplasty. The 2014 European Society of Cardiology (ESC)/European Association for Cardio-Thoracic Surgery (EACTS) guidelines recommended this approach for myocardial revascularization [108, 109].

Stent Angioplasty

This section will discuss the progress of stent angioplasty from the bare metal stents to newer stent platforms over the last two decades for improving healing and patient outcomes.

Table 6.2 US FDA-approved drug-eluting/drug-coated balloon catheter devices (DEB/DCB) for peripheral artery disease (PAD) [99–102]

Sr. no.	Balloon catheter	Year approved	Drug	Dose	Polymer
1	C.R. Bard's Lutonix 035 DCB catheter	2014	Paclitaxel	2 μg/ mm²	Specialized nonpolymer formulation with polysorbate, sorbitol as inactive ingredient
2	IN.PACT™ Admiral™ paclitaxel-coated percutaneous transluminal angioplasty (PTA) balloon catheter, Medtronic Inc.	2014	Paclitaxel	3.5 μg/ mm²	Proprietary FreePac™ coating solution of hydrophilic excipient (urea)
3	Stellarex OTW drug-coated balloon PTA catheter	2017	Paclitaxel	2 μg/ mm²	Proprietary coating of hydrophilic polymer excipient (polyethylene glycol 8000)

Bare Metal Stent (BMS) Angioplasty

Early limitations of balloon catheters encouraged the development of bare metal stents that enabled the widespread utilization of percutaneous coronary intervention (PCI) therapy worldwide. The biggest advantage of a bare metal stent over the balloon catheter was its mechanical strength that helped to overcome the effect of vessel recoil and constrictive remodelling and a reduction in the rate of restenosis [110]. Consequently, the design and development of new bare metal stent platforms had arrived with advanced medical therapies including dual antiplatelet therapy which had previously been limited early stent thrombosis and bleeding complications associated with thrombolytic therapy. The first licensed bare metal stent was made from 316L stainless steel which was subsequently replaced by an improved cobalt chromium metal alloy by Guidant Corporation. The advantage of this Guidant stent is its capability to produce a lower strut thickness with increased radial strength. Stent strut thickness is a key factor that plays an important role to reduce restenosis. The results of the ISAR-STEREO-2 clinical trial on a total of 611 patients indicated that the incidence of angiographic restenosis was 17.9% in the thin-strut stent group (50 micron) and was 31.4% in the thick-strut stent group (140 micron) [111]. Table 6.5 shows that stents are now available with different metal alloys, polymers and drugs with different strut thicknesses [112].

As a result of research and development of various metal alloys in the selection of modern stent materials, three basic properties have been taken into account. They are elasticity or plasticity for expansion, rigidity for the control of dilatation and resistance to elastic recoil. As shown in Table 6.5, various materials have been utilized for balloon expandable stents. They include 316L stainless steel, tantalum, martensitic nitinol, polymers, cobalt alloy, cobalt chromium alloy, hybrid tantalum with stainless steel, hybrid platinum with stainless steel and platinum chromium

Table 6.3 Drug-coated balloon (DCB) catheter devices available for percutaneous coronary intervention (PCI) [108]

Device	Company	Drug	Dose	Excipient
PACCOCATH®	Bayer, Germany	Paclitaxel	$3 \mu g/mm^2$	Iopromide
SeQuent® Please Neo	B. Braun Melsungen, Germany	Paclitaxel	$3 \mu g/mm^2$	Iopromide
DIOR I	Eurocor, Germany	Paclitaxel	$3 \mu g/mm^2$	Shelloic acid
DIOR II	Eurocor, Germany	Paclitaxel	$3 \mu g/mm^2$	Shelloic acid
Biostream	Biosensors International Group, Ltd., Switzerland	Paclitaxel	$3 \mu g/mm^2$	Shellac
Agent®	Boston Scientific, USA	Paclitaxel	$2 \mu g/mm^2$	Citrate ester
Essentia®	iVascular, S.L.U., Spain	Paclitaxel	$3 \mu g/mm^2$	Organic ester
IN-PACT Falcon™	Medtronic, USA	Paclitaxel	$3 \mu g/mm^2$	Urea
Genie™	Acrostak, Switzerland	Nanoporous	$10 \mu mol/L$	None
Pantera Lux®	Biotronik, Switzerland	Paclitaxel	$3 \mu g/mm^2$	Butyryl-tri-hexyl citrate
Elutax®	Aachen Resonance, Germany	Paclitaxel	$2 \mu g/mm^2$	Dextrane
Danubio®	Minvasys, France	Paclitaxel	$2.5 \mu g/mm^2$	Butyryl-tri-hexyl citrate
RESTORE® DEB	Cardionovum, Germany	Paclitaxel	$3 \mu g/mm^2$	Safepax
Protégé® and Protégé® NC	Blue Medical, Netherlands	Paclitaxel	$3 \mu g/mm^2$	Butyryl-tri-hexyl citrate
Virtue® DCB	Caliber Therapeutics, Inc., USA	Sirolimus nanoparticles	$3 \mu g/mm^2$	Porous balloon
Solution® DCB	M.A. Med Alliance SA, Switzerland	Sirolimus nanoparticles	$1 \mu g/mm^2$	Cell adherence technology (CAT)
MagicTouch™ Xtreme Touch™ DCB	Concept Medical, Surat, India	Sirolimus + nanocarriers	$1.27 \mu g/mm^2$ $3 \mu g/mm^2$	Phospholipid
Kanshas DCB	Terumo Corporation	Paclitaxel + Unicoat™ microcrystal coating	$3.2 \mu g/mm^2$	L-Serine Ethyl Ester HCl

alloy. For the development of self-expanding stents, alternate metals such as nickel titanium, nitinol, cobalt alloy, as well as novel biodegradable magnesium, iron (Fe), zinc (Zn) and their alloys have been used [113–116]. As thinner struts have been developed, the additional property of radio-opacity has become a major concern. Various methods, such as gold coating, radio-opaque dye coating or attachment of

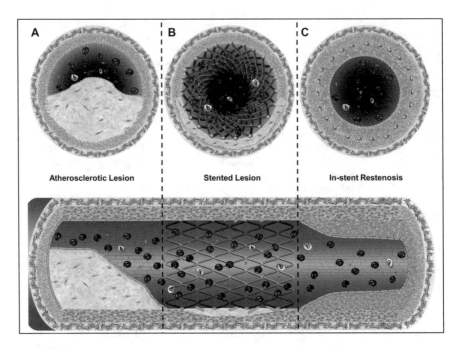

Fig. 6.5 Progression of in-stent restenosis. Cross-sectional and longitudinal views of an artery depicting the chronological progression of in-stent restenosis. (**a**) Obstructive atheromatous plaque causing flow-limiting stenosis of the arterial lumen with reduced luminal diameter. (**b**) After percutaneous endoluminal stenting which restores the native vessel diameter by compressing the atheromatous plaque into the vessel wall with resultant denudation of the endothelial layer. (**c**) In-stent restenosis after inappropriate neointimal hyperplasia in response to percutaneous stent insertion resulting in recurrence of flow-limiting stenosis [119]

a radio-opaque marker at the proximal and distal ends of the stent have been utilized [117]. One additional parameter that has been considered during metal stent development is the need to reduce the amount of metal ion release from the metal alloy, specifically nickel ions, after implantation. Strategies to reduce the release of metal ions have included using a nickel-free alloy or by using various coatings through chemical or physical vapour deposition methods such as diamond-like carbon coating, silicon carbide coating, carbon coating, titanium oxide coating, titanium-nitride-oxide coating or an iridium oxide coating [118].

Despite these advanced developments in bare metal stent technology, neointimal accumulation of plaque has remained the major limitation of bare metal stents, resulting in the development of in-stent restenosis (ISR) in 20–30% of cases as shown in Fig. 6.5. This restenosis limitation of the bare metal stent (BMS) has been referred to as the "Achilles' heel" of PCI and has led to the development of the first-generation drug-eluting stent system [109, 119].

Table 6.4 Three generations of drug-eluting stents (DES) [120, 121]

Generation	Device examples	Specifications	Advantage and limitations
First-generation DES	Cypher, Taxus	Drug: sirolimus, paclitaxel Platform: stainless steel, slotted tube design Polymer: durable	Superior to BMS in reducing the magnitude of neointimal proliferation and clinical restenosis Limitation: late stent thrombosis is more likely to occur with these stents
Second-generation DES	Endeavor, Xience V	Drug: zotarolimus, everolimus Platform: cobalt chromium, thin-strut stents Polymer: persistent	Superior to first-generation DES, exhibiting lower thrombosis rates Limitation: drugs delayed re-endothelialization
Newer generation DES	Axxess Stent, Nevo Stent	Drug: biolimus, sirolimus, everolimus Platform: platinum chromium, nickel-titanium Polymer: bioresorbable polymer coating, polymer free coating	Superior biocompatibility and controlled release drug profile

First-Generation Drug-Eluting Stent (DES) Angioplasty

Limitation of restenosis of the bare metal stent (BMS) led to the development of a drug-eluting stent (DES) which involved the controlled release of antiproliferative drugs incorporated within a polymer coating. At the same time as the development of bare metal stents and medical therapies, an early generation DES was developed that released sirolimus (e.g. the Cypher stent) or paclitaxel (e.g. the Taxus stent) from the relatively thick struts (120–140 μm) of a stainless steel stent platform coated with a polymer. To facilitate the controlled release of the drug, DES's utilized permanent synthetic polymer coating materials, known as biostable polymers, such as polyethylene-co-vinyl acetate, poly-n-butyl methacrylate and the tri-block copolymer poly(styrene-b-isobutylene-b-styrene). Tables 6.4 and 6.5 give an overview of all three generations of drug-eluting stents and list some of the US FDA-approved stents with their unique structural characteristics [120].

The first-generation DES was successful in reducing the angiographic and clinical restenosis by 50–70% compared to the bare metal stents, but it also increased the risk of late and very late stent thrombosis. As a result, the use of the first-generation DES was limited to certain conditions [122]. According to the research reported by Bønaa et al., the 6-year rate of repeat revascularization was less than about 16.5% in the drug-eluting stent group compared to 19.8% in the bare metal stent group ($p < 0.001$) [123]. Meta-analysis of a clinical trial comparing the 5-year follow-up

Table 6.5 Drug-eluting stents with durable or biodegradable polymer coatings [112, 121]

Sr. no.	Device	Platform	Drug	Strut thickness	Polymer type	Polymer material	Coating distribution	Polymer thickness	Additional coating
1	Taxus	SS	Paclitaxel	132 µm	Durable	SIBS	Circumferential	22	–
2	Cypher	SS	Sirolimus	140 µm	Durable	PEVA/PBMA	Circumferential	13	–
3	BioMatrix Nobori	SS	Biolimus	120 µm	Biodegradable	PDLLA	Abluminal	10	–
4	Endeavor	CoCr	Zotarolimus	91 µm	Durable	MPC/LMA/HPMA/3-MPMA	Circumferential	6	–
5	Yukon PC	SS	Sirolimus	87 µm	Biodegradable	PDLLA	Circumferential	5	–
6	Xience Promus	CoCr PtCr	Everolimus	81 µm	Durable	PBMA/PVDF-HFP	Circumferential	8	–
7	Resolute	CoCr	Zotarolimus	91 µm	Durable	PBMA/PHMA/PVP/PVA	Circumferential	6	–
8	Synergy	PtCr	Everolimus	74 µm	Biodegradable	PLGA	Abluminal	4	–
9	Orsiro	CoCr	Sirolimus	60 µm	Biodegradable	PLLA	Circumferential	7	Silicon carbide
10	DESyne	CoCr	Novolimus	81 µm	Biodegradable	PLLA	Circumferential	<3	–
11	Combo	SS	Sirolimus	100 µm	Biodegradable	PDLLA/PLGA	Abluminal	5	Anti-CD34 antibodies
12	Mistent	CoCr	Sirolimus	64 µm	Biodegradable	PLGA	Circumferential	10	–
13	Ultimaster	CoCr	Sirolimus	80 µm	Biodegradable	PDLLA-PCL	Abluminal	15	–

SS stainless steel, *CoCr* cobalt chromium, *PtCr* platinum chromium, *SIBS* poly(styrene-b-isobutylene-b-styrene), *PEVA* poly-ethylene-co-vinyl acetate, *PBMA* poly n-butyl methacrylate, *PVDF-HFP* co-polymer of vinylidene fluoride and hexafluoropropylene, *MPC* methacryloyloxyethyl phosphorylcholine, *LMA* lauryl methacrylate, *HPMA* hydroxypropyl methacrylate, *3-MPMA* trimethoxysilylpropyl methacrylate, *PVP* polyvinyl pyrrolidinone, *PHMA* polyhexyl methacrylate, *PVA* polyvinyl acetate, *PLGA* poly-lactic co-glycolic acid, *PLLA* poly-L-lactic acid, *PDLLA* poly-D, L-lactic acid

Table 6.6 First- and second-generation drug-eluting stent platforms showing year of FDA approval [121, 126–128]

Stent	Year	Manufacturer	Stent alloy	Drug
Cypher	2003	Cordis	Stainless steel	Sirolimus
Taxus	2004	Boston Scientific	Stainless steel	Paclitaxel
Xience V	2007	Abbott Vascular	Cobalt chromium	Everolimus
Promus	2008	Boston Scientific	Cobalt chromium	Everolimus
Endeavor	2008	Medtronic	Cobalt chromium	Zotarolimus
Xience Prime	2011	Abbott Vascular	Cobalt chromium	Everolimus
Promus Element	2011	Boston Scientific	Platinum chromium	Everolimus
Taxus Ion	2011	Boston Scientific	Cobalt chromium	Paclitaxel
Resolute	2012	Medtronic	Cobalt chromium	Zotarolimus
Promus Premier	2013	Boston Scientific	Platinum chromium	Everolimus
Synergy	2015	Boston Scientific	Platinum chromium	Everolimus
Resolute Onyx	2017	Medtronic	Platinum iridium core-cobalt alloy shell	Zotarolimus
EluNIR	2017	Medinol Ltd (Cordis)	Cobalt chromium	Ridaforolimus
ORSIRO	2019	Biotronik, Inc.	Cobalt chromium	Sirolimus

of the first-generation DES and bare metal stent implantations in 1414 patients showed a significant reduction in target vessel revascularization (TVR) (8.7% vs 14.8%), but an increase in very late stent thrombosis in the first-generation DES (3.0%) compared to the bare metal stent group (1.0%) [124].

Second-Generation Drug-Eluting Stent (DES) Angioplasty

The US FDA approved several second-generation drug-eluting stent devices, such as zotarolimus-eluting (e.g. Resolute), everolimus-eluting (e.g. Xience V) and ridaforolimus-eluting (e.g. EluNIR) stents (ZES, EES and RES) [125–127]. Table 6.6 mentions the first- and second-generation DES's along with the FDA year of approval. The newer DES stents have a platform of a cobalt chromium or platinum chromium alloy and are thinner, easier to deliver and are more biocompatible than the first-generation DES's as shown in Table 6.5. Advanced biostable and biodegradable polymers with advanced design features and metal alloys have been utilized in the second-generation DES to limit polymer-induced in-stent restenosis (ISR) and stent thrombosis (ST). Currently, drug-eluting stents are recommended over bare metal stents for any PCI, irrespective of the clinical presentation, lesion type, planned noncardiac surgery, anticipated duration of dual antiplatelet therapy (DAPT), concomitant anticoagulant therapy or radial access [122].

Drugs Used in DES's

There are several limus family members, such as everolimus, zotarolimus, biolimus A9, tacrolimus, novolimus and pimecrolimus, that have been researched for drug-eluting stent applications and taxus family members, such as cytotoxic paclitaxel, that have been used for PCI applications. Sirolimus, also called rapamycin, is a macrocyclic lactone, a chemical derivative of soil microorganisms. Sirolimus and other members of the limus family are active pharmaceutical ingredients (APIs) that inhibit the cell cycle progression between the late "Growth1" to "Synthesis" phase and thus prevent the proliferation and migration of vascular smooth muscle cells that is known to induce neointimal and restenosis development as shown in Fig. 6.6 [129, 130]. Pimecrolimus and tacrolimus are calcineurin inhibitors. At the cellular level, these drugs bind to the FK-binding protein 12 and subsequently inhibit the mammalian target of rapamycin (mTOR) which leads to increased tissue factor expression. Zotarolimus and everolimus are analogues of sirolimus and have similar immunosuppressant properties like sirolimus, but they have enhanced lipophilic properties due to their high log P values that prevent drug loss through blood flow.

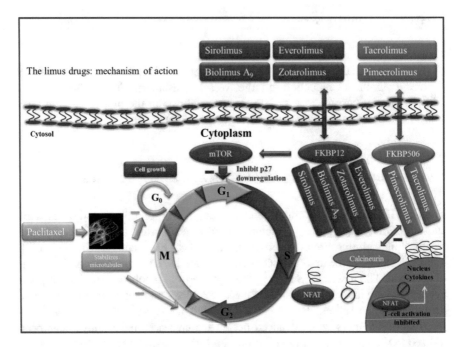

Fig. 6.6 Mechanisms of action of sirolimus, everolimus, biolimus A9, zotarolimus, tacrolimus and pimecrolimus [130] *FKBP* FK binding protein, *G* growth, *M* mitosis, *S* synthesis, *NFAT* nuclear factor of activated T cells, *mTOR* mammalian target of rapamycin

Note that tacrolimus has a less inhibitory effect on smooth muscle cell proliferation compared to sirolimus [131–134].

Paclitaxel is a potent cytostatic agent which inhibits cell proliferation and migration by disrupting the delivery of cellular microtubules. Paclitaxel interupts the cell cycle through stabilizing longer microtubules during mitosis by preventing the transition from the "Growth2" to the "Mitosis" phase which leads to the inhibition of smooth muscle cell proliferation and neointimal formation. The clinical trial KAMIR compared the paclitaxel- versus sirolimus-eluting stents for the treatment of STEMI patients and concluded that Paclitaxel is the preferred drug for angioplasty balloons due to its lipophilic properties that enable the drug-coated balloon to deliver the drug to the vessel wall during a shorter contact time. However, it was found that sirolimus was superior to paclitaxel with regards to the occurrence of a major adverse cardiac event (MACE) and target lesion revascularization (TLR) [135, 136].

Polymers Used in DES's

During the stages of DES development, one of the major factors affecting the in-vivo performance of the stent is the biocompatibility and in-vivo behaviour of the polymer when releasing the drug [137]. The purpose of the polymer is to modulate the elution of the drug into the arterial wall at the site of implantation. Over the years, polymers such as biostable/durable permanent polymers for the first-generation stents and biodegradable polymers for the second-generation stents have been optimized to provide a nonthrombotic, noninflammatory, nontoxic and re-endothelialisable performance [138].

Biostable or durable polymers, such as nonerodable polyethylene-co-vinyl acetate (PEVA) and poly-n-butyl methacrylate (PBMA) (Cypher Stent) and soft elastomeric polymers such as poly(styrene-bisobutylene-b-styrene) (SIBS) (Taxus, Promus and ION Stent), phosphorylcholine (ZoMaxx, Endeavor ZES), Biolinx polymer (Endeavor Resolute), polyvinylidene fluoride co-hexafluoropropylene (PVDF-HFP), poly-n-butyl methacrylate (Xience V and Promus stent), methacrylated phosphorylcholine based (PC) (Endeavor stent), a combination of poly(butyl methacrylate-co-vinyl acetate) (C10), poly(hexyl methacrylate-co-vinyl pyrrolidone-co-vinyl acetate) (C19) and poly(vinyl pyrrrolidone) (PVP) (Resolute stent), have to date been used on devices approved by the US FDA [120, 121, 139, 140].

Some of the latest generation of stents with a biodegradable polymer coating have shown improvement in the efficacy of DES's with a lower rate of late in-stent thrombosis. The presence of late in-stent thrombosis for durable polymer-coated stents depends on the polymer as well as other stent-thrombosis factors, as mentioned in Fig. 6.3 [120, 141–143]. Table 6.7 lists the metallic stents with a biodegradable polymer coating or a polymer free technology [120, 121, 144–146].

Table 6.7 Metallic drug coated stents using a biodegradable polymer or polymer free coating technology together with an active pharmaceutical ingredient (API) [120, 121, 140, 145–150]

Polymer/technology to coat API	Platform alloy	Stent	API
Polylactic acid (PLA)	Nickel–titanium (Nitinol)	Axxess (Devax Inc.)	Biolimus A9
	L605 cobalt–chromium (Co-Cr)	Custom NX, Xtent (Xtent)	Biolimus A9
	316l stainless steel (316L SS)	BioMatrix (biosensors)	Biolimus A9
	316L SS	Nobori (Terumo)	Biolimus A9
	Nickel–titanium (Nitinol)	Axxess (Devax Inc.)	Biolimus A9
	L605 Co-Cr	Lumeno-Alpha (Cordis)	Biolimus A9
	L605 Co-Cr	Elixir Myolimus (Elixir Medical)	Myolimus
	316L SS with drug-polymer microdot coating	JACTAX HD (Boston Scientific) (134)	Paclitaxel
	316L SS	Excel stent (JW Medical System)	Sirolimus
	L605 Co-Cr	DESny BD (Elixir)	Novolimus
	316L SS	Champion (Boston Scientific Corp.)	Everolimus
Poly L-lactic acid (PLLA)	L605 Co-Cr	ORSIRO (Biotronik)	Sirolimus
Bioabsorbable, polylactide-co-glycolide (PLGA)	L605 Co-Cr with unique reservoir for drug-polymer	NEVO (Cordis, Johnson & Johnson)	Sirolimus
	316L SS	CORACTO (ALVIMEDICA)	Sirolimus
	Platinum chromium	SYNERGY (Boston Scientific)	Everolimus
	L-605 Co-Cr with micropore	CoStar stent (Conor MedSystems)	Paclitaxel
PLGA + PLLA	L-605 Co-Cr	BioMime (Meril Life Sciences)	Sirolimus
	Nitinol	Cardiomind (CardioMind Inc.)	Sirolimus
Bioabsorbable polymer, containing poly-L-lactide, polyvinyl pyrrolidone, polylactide-co-caprolactone and polylactide-co-glycolide	L-605 Co-Cr alloy with layered coating for controlled release	Supralimus-core (Sahajanand Medical)	Sirolimus
Poly (DL-lactide-co-caprolactone)	L-605 Co-Cr with no coating at strut curvature (hinges)	Ultimaster (Terumo)	Sirolimus
Polymer free/passive coating			
Selectively microstructured surface	L605 Co-Cr	Lumeno Free (Cordis)	Biolimus A9

(continued)

Table 6.7 (continued)

Polymer/technology to coat API	Platform alloy	Stent	API
Microstructured abluminal stent surface	316L SS	Biofreedom (Biosensors International)	Biolimus A9
Abluminal spray coated on synthetic glycocalix substrate	316L SS	Axxion (Biosensors International)	Paclitaxel
Direct drug coating	316L SS	Achieve (Cook Inc.)	Paclitaxel
	316L SS	Supra-G (Cook Inc.)	Paclitaxel
	316L SS	V-Flex Plus (Cook Inc.)	Paclitaxel
Abluminal coating with crystallized drug (microdrop spray crystallization process)	L605 Co-Cr	Amazonia PAX (Minvasys)	Paclitaxel
Microporous stent surface with top coat of shellac resin	316L SS	Yukon Choice (Translumina GmbH)	Sirolimus
Drug loading with micropores	316L SS	Yinyi (Liaoning Biomedical Materials)	Paclitaxel
Adluminal bio-inducer surface coating and drug reservoir	L605 Co-Cr	Cre8 (Alvimedica)	Amphilimus
Adluminal integral carbofilm coating and drug reservoir	316L SS	Janus (Sorin Biomedica)	Tacrolimus
Abluminal nanoporous cavities as drug reservoir	316L SS	Nano + (Lepu Medical)	Sirolimus
Diffusion from stent core	Co-Cr outer cover strut with hollow core for drug	Polymer-free drug-filled stent (Medtronic)	Sirolimus
Micropores with shellac resin	316L SS	Dual-DES	Sirolimus and antioxidant probucol
Nanothin-microporous hydroxyapatite coating	316L SS	Vestasync (MIV Therapeutics)	Sirolimus

Over the years, there have been developments such as advanced drug elution profiles, thinner and stronger strut profiles, improved biocompatible metal alloy platforms, advanced durable and biodegradable polymer coatings and advanced post-procedural medical therapies for the second generation of drug-eluting stents. These developments have resulted in a significant reduction in in-stent restenosis (ISR), primarily due to a lower incidence of neointimal hyperplasia [151, 152]. Two large-scale comprehensive clinical trials which included 12,866 and 18,334 patients utilized network meta-analysis of the data and concluded the superiority of the second-generation DES over bare metal stents, and the first-generation DES in terms of safety and efficacy. However, the very long-term clinical implications, such as very late stent thrombosis, are still a major concern for the patient population. Second-generation DES's have the side effects, such as delayed re-endothelialization, neoatherosclerosis, medical necrosis and chronic inflammation. The use of dual antiplatelet therapy (DAPT) or anticoagulant medical therapy is prescribed to minimize these side effects which also means that there is still scope for improvement in the stent design and materials used [122, 153–159].

Although it has been proven that the second-generation drug-eluting stents have the ability to overcome the limitations of bare metal stents and first-generation DES's, in order to make further improvements, we are challenged by the performance of the second-generation DES drugs. Restenosis can be caused by an overreaction of the wound-healing response at the site of the stent injured vessel, while re-endothelialization is essential for normal wound healing. DES drugs, on the one hand, inhibit overreaction and restenosis, but on the other hand, by impairing the healing process of the injured arterial wall, these drugs result in delayed re-endothelialization and the formation of incompetent endothelium in terms of integrity and function. Such incompetent endothelium after DES therapy leads to accelerated and more frequent in-stent neoatherosclerosis [160, 161].

Current drug-eluting stents, as described above, depend on one of two mechanisms:

1. A passive coating to prevent cellular adhesion and subsequent inflammation
2. The use of a cytotoxic and antiproliferative drug

Both strategies increase the healing time due to their effect on delaying the re-endothelialization process and ultimately slowing healing. Delayed healing is the major risk associated with thrombosis and late stent restenosis [162, 163]. It has also been proven that the drugs loaded on drug-eluting stents inhibit proliferation, migration, differentiation and the endothelial nitric oxide synthase (eNOS) production in human endothelial progenitor cells [164].

Drug-Eluting Polymeric Stents

A novel development in the field of polymeric stents is the use of bioresorbable polymers instead of a permanent metal alloy. The advantage of these polymeric stents is that they absorb at the site of implantation within 6–24 months after implantation and once resorbed, it leaves no residue behind that will cause an inflammatory response. Thus, it preserves the vessel biomechanics, such as the vessel vasomotion. Due to the limited radial compression resistance and polymer-induced inflammation, these resorbable polymeric stents have a tendency to experience in-stent thrombosis. They are currently being evaluated clinically [165–167]. Table 6.8 mentions various polymers that are being used as a bioresorbable stent platform with their respective resorption time.

Table 6.8 List of commercially available and investigated bioresorbable scaffolds and specifications [165, 168–170]

Scaffold name	Strut material	Coating material	Eluted drug	Strut thickness	Radio-opacity	Resorption (months)	Crossing profile (mm)	Current status
Igaki-Tamai BRS	PLLA[a]	None	None	170	Gold	24-36	–	CE for PAD[a]
Absorb BVS[a] 1.0	PLLA	PDLLA[a]	EVL[a]	156	Pt[a]	18-24	1.4	Discontinued
Absorb BVS 1.1	PLLA	PDLLA	EVL[a]	156	Pt[a]	24-48	1.4	CE mark
Absorb new generation	PLLA	PDLLA	EVL[a]	<100	–	–	–	–
DESolve	PLLA	None	MYL[a]	150	Metallic	12-24	1.5	CE mark
DESolve 100	PLLA	PLLA	NVL[a]	100	–	24	–	CE mark
Reva scaffold	PTD-PC[a]	None	None	200	ROS[a]	24	0.1.8	Discontinued
ReZolve scaffold	PTD-PC	None	SRL[a]	115-230	ROS[a]	4-6	1.8	Clinical trial
ReZolve[2] scaffold	PTD-PC	None	SRL[a]	100	ROS[a]	48	1.5	Clinical trial
Fantom	PTD-PC	–	SRL[a]	125	–	36	–	Clinical trial
Ideal Biostent	Polymer salicylate	Salicylate	SRL[a]	175	None	>12	1.5-1.7	Clinical trial
Art18z BRS	PDLLA	None	None	170	None	3-6	6-Fr	Clinical trial
Amaranth	Semicrystalline polylactide		None	90-150	None	3-6	6-Fr	Clinical trial
Xinsorb BRS	PLLA	PDLLA	SRL[a]	160	–	24-36	–	Clinical trial
Acute BRS	PLCL,[a] PDLA, PLLA	–	SRL[a]	150	–	–	–	–
MeRes	PLLA	PDLLA	MRL[a]	100	–	24	–	Clinical trial
Fades	PLGA[a] & Mg[a]	–	–	–	–	6	–	–
Mirage bioresorbable micro-fibre scaffold	PLLA	–	SRL[a]	125-150	–	14	0.44"-0.058"	Clinical trial
AMS-1	Mg alloy	None	None	165	None	1	1.4	Discontinued
Dreams[a] 1.0	Mg with rare metal	PLGA	PXL[a]	120	None	3	1.2	Clinical trial

(continued)

Table 6.8 (continued)

Scaffold name	Strut material	Coating material	Eluted drug	Strut thickness	Radio-opacity	Resorption (months)	Crossing profile (mm)	Current status
Dreams[a] 2.0	Mg with rare metal	PLLA	SRL[a]	150	TNT[a] marker	12	1.75	Clinical trial
Unity BDS	Mg/PLLA	–	SRL[a]	–	–	12	–	Preclinical test
Dreams 3.0	Mg (WE43)	PLLA	SRL[a]	99,117,147	Yes	12	–	Preclinical test
NOR-I	Iron	NA	NA	91	–	41 mg Fe/month	–	Preclinical test
IBS	Iron	Zinc+ PLLA	SRL[a]	70	–	13	1.04	Preclinical test

[a]*PLLA* poly-L-lactic acid, *PDLLA* poly-D,L-lactic acid, *BVS* bioresorbable vascular scaffold, *PTD-PC* poly-tyrosine-derived polycarbonate, *PLCL* poly-L-lactide-co-ε-caprolactone, *PLGA* poly-lactide-co-glycolide, *DREAMS* drug-eluting absorbable metallic stents, *EVL* everolimus, *MYL* myolimus, *NVL* novolimus, *SRL* sirolimus, *PXL* paclitaxel, *MRL* merilimus, *Mg* magnesium, *Pt* Platinum, *ROS* radio-opaque scaffold, *PAD* peripheral artery disease, *TNT* tantalum

Recent Advances in Angioplasty Devices

Recently, scientists are trying to improve DES technology by studying different approaches like innovative drug delivery or modifications to the stent platforms as listed in Tables 6.7, 6.8 and 6.9 [171, 172]. Some of the approaches have included the following:

1. Delivery of the drug on only one side or portion of that side of the stent's surface. This can be achieved by coating one side only or using microdots or reservoirs on the stent surface that leave one aspect of the stent surface (luminal or peripheral) as a bare metal surface.
2. The use of biomolecules or antibodies to mimic the natural tissue in order to diminish the inflammatory response or to encourage antibodies to capture circulating endothelial progenitor cells to promote healing.
3. Use a polymer free technology such as a hollow stent strut to be filled with the drug in order to have more controlled release of the drug without a polymer coating. Polymer-free stents having a microporous stent surface to hold the drug without the risk of a polymer-induced inflammatory reaction as mentioned in Table 6.7. But clinical trials have failed to prove any superiority over drug-polymer-eluting stents [148, 173].
4. Use of a closed cell stent design instead of an open cell to limit the size of the injury site exposed to blood and promote re-endothelialization [174–176].

Thia chapter has presented the entire timeline for the development of angioplasty technologies starting from the first balloon angioplasty in 1977 to the current technologies that are either in commercial production and clinical use or they are under investigation as mentioned in Tables 6.7, 6.8 and 6.9. There is still much room for improvement and more clinical data is requied for these current techologies and investigation approaches such as biomolecules, surface modifications, antibodies and peptide applications. There are some key factors, such as radial strength, bioresorption, stent geometry, radio-opacity, novel drug molecules and their release profile, that need to be evaluated and assessed while developing more efficient and safer angioplasty devices for the future.

Table 6.9 List of current technologies, coatings and biomolecules under study to improve stent performance [145, 147, 176–187]

Sr. no.	Technology/coating/biomolecules to improve stent efficacy	Type/category/stent platform	Mechanism/action	Effect/result
1	Heparin-functionalized coating	Sulphated glycosaminoglycan	Reaction with antithrombin III, inhibited human umbilical artery smooth muscle cell (HUASMC) adhesion and proliferation	Anticoagulant, anti-inflammatory, platelet and smooth muscle cell (SMC) inhibitor
2	Fucoidan	Sulphated polysaccharide	Recruitment of endothelial progenitor cells (EPCs), reduce SMCs proliferation	Prevents neointimal hyperplasia
3	Laminins	Glycoproteins	Mediate both cell-to-cell and cell-to-extracellular matrix adhesion, anticoagulant property	Enhance cell differentiation and proliferation of endothelial cells (ECs)
4	Chondroitin sulphate	Sulphated polysaccharide	Electrostatic repulsion towards fibrinogen	Hemocompatible, enhance the resistance of vascular cell apoptosis
5	Hyaluronic acid	Negatively charged nonsulphated polysaccharide	Cell attachment and signalling through interacting with cell surface receptors and reduce adhesion of platelets	Platelet inhibition
6	Fibronectin	Major component in ECM	Interaction with α5β1 transmembrane integrin receptor of EC cells	Promote the attachment, spreading and differentiation of ECs
7	Vascular endothelial growth factor (VEGF)	VEGF-functionalized on titanium substrate	Induce the differentiation of hMSCs into endothelial cells	Promote re-endothelialization
8	VEGF + hepatocyte growth factor (HGF) by umbilical cord blood-derived mesenchymal stem cell (UCB-MSC)-seeded stent	UCB-MSC secreted growth factors on stent seeded for 7 days	Reduced restenosis within the stent and induced natural re-endothelialization	Improved healing
9	Gallic acid	Natural plant phenol molecule	Antioxidant, anti-inflammatory property	Induce SMC death, promote EC growth
10	CD34 + antibody	API coating on stainless steel stent (Genous, OrbusNeich)	Bind to endothelial progenitor cells (EPCs)	Reduce ST and restenosis

11	Heparin/poly-l-lysine microspheres (MS) immobilized on dopamine-coated stent	MS form covalent bond with dopamine through Schiff base and/or Michael addition reaction	Improved cytocompatibility, blood compatibility	Accelerate endothelialization
12	Polydopamine (PDA) functionalized titanium dioxide nanotubes (TiO_2 NTs)	Topography modification with bio-inspired coating material	Anti-inflammatory, enhance EC adhesion	Reduce SMC adhesion and proliferation
13	Dopamine-conjugated hyaluronic acid with sirolimus	Coated with in poly(d,l-lactide) on stent	Suppressive effects on platelet adhesion and activation	Maintained the EC viability and proliferation
14	Titanium nitride oxide	Passive coating on stainless steel stent (Titan-2)	Inhibit platelet aggregation, minimize fibrin deposition, reduce inflammation	Promote healing
15	Polyzene F	NanoThin Polyzene-F polymer coating on CoCr stent (Catania)	Anti-inflammatory, bacteria resistant and pro-healing qualities	Low surface thrombogenicity
16	Paclitaxel and pimecrolimus loaded in adjacent reservoirs on stent surface micropores	SymBio stent (Conor Medsystems)	Combine effect of both drugs	Failed to show superiority over current DES
17	Sirolimus and anti-CD34 antibody on SynBiosys™ (PLA) coating	Genous L605 Co-Cr stent (Orbus Neich)	Antiproliferation by sirolimus and EPC capture effect of anti-CD34	Improved healing compared to BMS
18	Vascular smooth muscle cell (VSMC)-like biomimetic surface patterns on stents using a femtosecond laser	316L stainless steel stent nano and micropattern	Promote adhesion, proliferation and migration of VSMC	Rapid re-endothelialization
19	Nitric oxide (NO) producing coating mimicking endothelium function	NO-catalytic bioactive coating to generate NO on 316L SS stent	Suppression of collagen-induced platelet activation and aggregation	Enhanced human umbilical vein endothelial cell (HUVEC) adhesion, proliferation and migration for re-endothelialization

(continued)

Table 6.9 (continued)

Sr. no.	Technology/coating/biomolecules to improve stent efficacy	Type/category/stent platform	Mechanism/action	Effect/result
20	Use of micronet around the stent to make it closed cell design	Use of mesh covering to minimize the open struts areas thus expose to blood flow	Eliminate post-procedural debris embolization, inhibit platelet activation	Prevent cardiac microvascular dysfunction or brain stroke in case of carotid stenting
21	Mesenchymal stem cells	Cell therapy for cardiac regeneration	Inhibit the vascular smooth muscle cell proliferation, enhance neovascularization	Rapid re-endothelialization
22	Controlled delivery of sirolimus from a bioactive polymer (accelerate™ AT)	Electrospray (ES) deposition techniques onto stent surface	Support endothelial cell growth	Promotes re-endothelialization and protection against in-stent restenosis and thrombosis

API active pharmaceutical ingredient, *ST* stent thrombosis, *ECs* endothelial cells, *CoCr* cobalt-chromium, *ECM* extracellular matrix

References

1. Lawlor M, Gersh B, Opie L, Gaziano T. Chapter 2 – The global perspective of ischemic heart disease. In: Chronic coronary artery disease. Philadelphia: Elsevier; 2018. p. 16–31.
2. Murphy SL, Xu J, Kochanek KD, Curtin SC, Arias E. Deaths: final data for 2015. Natl Vital Stat Rep. 2017 Nov;66(6):1–75.
3. Benjamin EJ, Paul M, Alvaro A, Bittencourt MS, Callaway CW, Carson AP, et al. Heart disease and stroke statistics—2019 update: a report from the American Heart Association. Circulation. 2019;139(10):e56–e528.
4. GBD 2013 Mortality and Causes of Death Collaborators. Global, regional, and national age–sex specific all-cause and cause-specific mortality for 240 causes of death, 1990–2013: a systematic analysis for the global burden of disease study 2013. Lancet. 2015;385(9963):117–71.
5. Kloner RA, Jennings RB. Consequences of brief ischemia: stunning, preconditioning, and their clinical implications: part 1. Circulation. 2001;104(24):2981.
6. Bonow RO, Mann DL, Zipes DP, Libby P. Braunwald's heart disease: a textbook of cardiovascular medicine. Philadelphia: Elsevier/Saunders; 2015.
7. Vise S. Cardiovascular disease: a costly burden for America, projections through 2035. Health Metr. 2017;2017: 5–14
8. Mozaffarian D, Benjamin EJ, Go AS, Arnett DK, Blaha MJ, Cushman M, et al. Heart disease and stroke statistics—2015 update: a report from the American Heart Association. Circulation. 2015;131(4):e29–e322.
9. Prescott E. Chapter 18 – Lifestyle interventions. In: de Lemos JA, Omland T, editors. Chronic coronary artery disease. Philadelphia: Elsevier; 2018. p. 250–69.
10. Lavie CJ, De Schutter A, Milani RV. Chapter 19 – Obesity and the obesity paradox. In: de Lemos JA, Omland T, editors. Chronic coronary artery disease. Philadelphia: Elsevier; 2018. p. 270–9.
11. Wilson PWF, O'Donnell CJ. Chapter 1 – Epidemiology of chronic coronary artery disease. In: Chronic coronary artery disease. Philadelphia: Elsevier; 2018. p. 1–15.
12. Kosiborod M, Arnold SV. Chapter 16 – Goals of therapy. In: de Lemos JA, Omland T, editors. Chronic coronary artery disease. Philadelphia: Elsevier; 2018. p. 227–33.
13. Libby P, Ridker PM, Hansson GK. Progress and challenges in translating the biology of atherosclerosis. Nature. 2011;473(7347):317–25.
14. Hansson GK. Mechanisms of disease: inflammation, atherosclerosis, and coronary artery disease. N Engl J Med. 2005;352(16):1685–95.

15. Spencer JH, Anderson SE, Lahm R, Iaizzo PA. The coronary vascular system and associated medical devices. In: Iaizzo PA, editor. Handbook of cardiac anatomy, physiology, and devices. Cham: Springer; 2015. p. 137–61.

16. Fortier A, Gullapalli V, Mirshams R. Review of biomechanical studies of arteries and their effect on stent performance. IJC Heart & Vessels. 2014;4:12–8.

17. Swirski FK, Libby P, Aikawa E, Alcaide P, Luscinskas FW, Weissleder R, et al. Ly-6C(hi) monocytes dominate hypercholesterolemia-associated monocytosis and give rise to macrophages in atheromata. J Clin Invest. 2006;117(1):195–205.

18. Chistiakov DA, Melnichenko AA, Myasoedova VA, Grechko AV, Orekhov AN. Mechanisms of foam cell formation in atherosclerosis. J Mol Med. 2017;95(11):1153–65.

19. Bobryshev YV. Monocyte recruitment and foam cell formation in atherosclerosis. Micron. 2006;37(3):208–22.

20. Bäck M, Hansson G. Chapter 4 – Basic mechanisms of atherosclerosis. In: de Lemos JA, Omland T, editors. Chronic coronary artery disease. Philadelphia: Elsevier; 2018. p. 45–54.

21. Borow KM, Nelson JR, Mason RP. Biologic plausibility, cellular effects, and molecular mechanisms of eicosapentaenoic acid (EPA) in atherosclerosis. Atherosclerosis. 2015;242(1):357–66.

22. Ma S, Fan L, Cao F. Combating cellular senescence by sirtuins: implications for atherosclerosis. Biochim Biophys Acta Mol Basis Dis. 2019;1865(7):1822–30.

23. Kolodgie FD, Burke AP, Farb A, Gold HK, Yuan J, Narula J, et al. The thin-cap fibroatheroma: a type of vulnerable plaque: the major precursor lesion to acute coronary syndromes. Curr Opin Cardiol. 2001 Sep;16(5):285–92.

24. Anlamlert W, Lenbury Y, Bell J. Modeling fibrous cap formation in atherosclerotic plaque development: stability and oscillatory behavior. Adv Differ Equ. 2017;2017(1):195.

25. Lopes J, Adiguzel E, Gu S, Liu S, Hou G, Heximer S, et al. Type VIII collagen mediates vessel wall remodeling after arterial injury and fibrous cap formation in atherosclerosis. Am J Pathol. 2013;182(6):2241–53.

26. Finn AV, Nakano M, Narula J, Kolodgie FD, Virmani R. Concept of vulnerable/unstable plaque. Arterioscler Thromb Vasc Biol. 2010;30(7):1282–92.

27. Yahagi K, Otsuka F, Sakakura K, Joner M, Virmani R. Native coronary artery and bypass graft atherosclerosis. In: Lanzer P, editor. PanVascular medicine. Berlin/Heidelberg: Springer; 2015. p. 273–301.

28. Kyavar M, Alemzadeh-Ansari M. Chapter 35 – Stable ischemic heart disease. In: Maleki M, Alizadehasl A, Haghjoo M, editors. Practical cardiology. St. Louis: Elsevier; 2018. p. 591–630.

29. Alexopoulos N, Raggi P. Calcification in atherosclerosis. Nat Rev Cardiol. 2009;6:681.

30. Veit S, Lima JA, Bluemke DA. Noninvasive imaging of atherosclerotic plaque progression. Circ Cardiovasc Imaging. 2015;8(7):e003316.

31. Insull W Jr. The pathology of atherosclerosis: plaque development and plaque responses to medical treatment. Am J Med. 2009;122(1):S3–S14.

32. Quillard T, Croce KJ. Pathobiology and mechanisms of atherosclerosis. In: Aikawa E, editor. Cardiovascular imaging: arterial and aortic valve inflammation and calcification. Cham: Springer; 2015. p. 3–38.

33. Nakahara T, Strauss HW. From inflammation to calcification in atherosclerosis. Eur J Nucl Med Mol Imaging. 2017;44(5):858–60.

34. Durham AL, Speer MY, Scatena M, Giachelli CM, Shanahan CM. Role of smooth muscle cells in vascular calcification: implications in atherosclerosis and arterial stiffness. Cardiovasc Res. 2018;114(4):590–600.

35. Doherty TM, Asotra K, Fitzpatrick LA, Qiao J, Wilkin DJ, Detrano RC, et al. Calcification in atherosclerosis: bone biology and chronic inflammation at the arterial crossroads. Proc Natl Acad Sci U S A. 2003;100(20):11201–6.

36. Weaver J. Insights into how calcium forms plaques in arteries pave the way for new treatments for heart disease. PLoS Biol. 2013;11(4):e1001533.

37. Cocker MS, McArdle B, Spence JD, Lum C, Hammond RR, Ongaro DC, et al. Imaging atherosclerosis with hybrid [18F]fluorodeoxyglucose positron emission tomography/computed tomography imaging: what leonardo da vinci could not see. J Nucl Cardiol. 2012;19(6):1211–25.

38. Albanese I, Khan K, Barratt B, Al Kindi H, Schwertani A. Atherosclerotic calcification: Wnt is the hint. J Am Heart Assoc. 2018;7:e007356.

39. Sedehi D, Cigarroa JE. Chapter 6 – Precipitants of myocardial ischemia. In: de Lemos JA, Omland T, editors. Chronic coronary artery disease. Philadelphia: Elsevier; 2018. p. 69–77.

40. Hamm CW, Bassand JP, Agewall S, Bax J, Boersma E, Bueno H, et al. ESC guidelines for the management of acute coronary syndromes in patients presenting without persistent ST-segment elevation: the Task Force for the management of acute coronary syndromes (ACS) in patients presenting without persistent ST-segment elevation of the European Society of Cardiology (ESC). Eur Heart J. 2011 Dec;32(23):2999–3054.

41. Nikus K, Birnbaum Y, Eskola M, Sclarovsky S, Zhong-Qun Z, Pahlm O. Updated electrocardiographic classification of acute coronary syndromes. Curr Cardiol Rev. 2014;10(3):229–36.

42. Amsterdam EA, Wenger NK, Brindis RG, Casey J, Donald E, Ganiats TG, Holmes J, David R, et al. 2014 AHA/ACC guideline for the management of patients with non-ST-elevation acute coronary syndromes: a report of the American College of Cardiology/American Heart Association Task Force on Practice Guidelines. J Am Coll Cardiol. 2014;64(24):e139.

43. Weber C, Noels H. Atherosclerosis: current pathogenesis and therapeutic options. Nat Med. 2011;17(11):1410–22.

44. Acute coronary syndrome – diagnosis of ACS; 2018. https://www.thrombosisadviser.com/acute-coronary-syndrome/#diagnosis-of-acs

45. Alemzadeh-Ansari M. Chapter 3 – Electrocardiography. In: Maleki M, Alizadehasl A, Haghjoo M, editors. Practical cardiology. St. Louis: Elsevier; 2018. p. 17–60.

46. Sahoo KP, Thakkar KH, Lin W, Chang P, Lee M. On the design of an efficient cardiac health monitoring system through combined analysis of ECG and SCG signals. Sensors. 2018;18(2):E379.

47. Camici PG, Rimoldi OE, Crea F. Chapter 5 – Coronary microvascular dysfunction. In: de Lemos JA, Omland T, editors. Chronic coronary artery disease. Philadelphia: Elsevier; 2018. p. 55–68.

48. Xu J, Lo S, Juergens CP, Leung DY. Assessing coronary microvascular dysfunction in ischaemic heart disease: little things can make a big difference. Heart Lung Circ. 2020;29(1):118–27.

49. Taqueti VR, Di Carli MF. Coronary microvascular disease pathogenic mechanisms and therapeutic options. J Am Coll Cardiol. 2018;72(21):2625.

50. De Bruyne B, Oldroyd KG, Pijls NHJ. Microvascular (dys)function and clinical outcome in stable coronary disease. J Am Coll Cardiol. 2016;67(10):1170.

51. Ford TJ, Corcoran D, Berry C. Stable coronary syndromes: pathophysiology, diagnostic advances and therapeutic need. Heart. 2018;104(4):284.

52. Camici PG, d'Amati G, Rimoldi O. Coronary microvascular dysfunction: mechanisms and functional assessment. Nat Rev Cardiol. 2015;12(1):48–62.

53. Crea F. Coronary microvascular dysfunction electronic resource. Milan: Springer; 2014.

54. Naderi S. Microvascular coronary dysfunction—an overview. Curr Atheroscler Rep. 2018;20(2):7.

55. Crea F, Lanza GA, Camici PG. CMD in the absence of myocardial diseases and obstructive CAD. In: Crea F, Lanza GA, Camici PG, editors. Coronary microvascular dysfunction. Milano: Springer; 2014. p. 75–114.

56. Crea F, Lanza GA, Camici PG. Mechanisms of coronary microvascular dysfunction. In: Crea F, Lanza GA, Camici PG, editors. Coronary microvascular dysfunction. Milano: Springer; 2014. p. 31–47.

57. Böse D, von Birgelen C, Zhou XY, Schmermund A, Philipp S, Sack S, et al. Impact of athero-sclerotic plaque composition on coronary microembolization during percutaneous coronary interventions. Basic Res Cardiol. 2008;103(6):587–97.

58. Juan-Carlos K, Filippo C, Gersh BJ, Camici PG. Reappraisal of ischemic heart disease. Circulation. 2018;138(14):1463–80.

59. Boden WE, O'Rourke RA, Teo KK, Hartigan PM, Maron DJ, Kostuk WJ, et al. Optimal medical therapy with or without PCI for stable coronary disease. N Engl J Med. 2007;356(15):1503–16.

60. Milo M, Nerla R, Tarzia P, Infusino F, Battipaglia I, Sestito A, et al. Coronary microvascular dysfunction after elective percutaneous coronary intervention: correlation with exercise stress test results. Int J Cardiol. 2013;168(1):121–5.

61. Testa L, Van Gaal WJ, Biondi Zoccai GG, Agostoni P, Latini RA, Bedogni F, et al. Myocardial infarction after percutaneous coronary intervention: a meta-analysis of troponin elevation applying the new universal definition. QJM. 2009 Jun;102(6):369–78.

62. Izzo P, Macchi A, De Gennaro L, Gaglione A, Di Biase M, Brunetti ND. Recurrent angina after coronary angioplasty: mechanisms, diagnostic and therapeutic options. Eur Heart J Acute Cardiovasc Care. 2012;1(2):158–69.

63. Enriquez JR, Parikh SV. Chapter 7 – History and physical examination. In: de Lemos JA, Omland T, editors. Chronic coronary artery disease. Philadelphia: Elsevier; 2018. p. 79–87.

64. Mancini GBJ, Gosselin G, Chow B, Kostuk W, Stone J, Yvorchuk KJ, et al. Canadian cardiovascular society guidelines for the diagnosis and management of stable ischemic heart disease. Can J Cardiol. 2014;30(8):837–49.

65. Akita CA, McGee SR. Bedside diagnosis of coronary artery disease: a systematic review. Am J Med. 2004;117(5):334–43.

66. Patel MR, Bailey SR, Bonow RO, Chambers CE, Chan PS, Dehmer GJ, et al. ACCF/SCAI/AATS/AHA/ASE/ASNC/HFSA/HRS/SCCM/SCCT/SCMR/STS 2012 appropriate use criteria for diagnostic catheterization: a report of the American College of Cardiology foundation appropriate use criteria task force, Society for Cardiovascular Angiography and Interventions, American Association for Thoracic Surgery, American Heart Association, American Society of Echocardiography, American Society of Nuclear Cardiology, Heart Failure Society of America, Heart Rhythm Society, Society of Critical Care Medicine, Society of Cardiovascular Computed Tomography, Society for Cardiovascular Magnetic Resonance, and Society of Thoracic Surgeons. J Am Coll Cardiol. 2012;59(22):1995–2027.

67. Thamwiwat A, Levine GN. Chapter 14 – Chest pains and angina. In: Levine GN, editor. Cardiology secrets. 5th ed. Philadelphia: Elsevier; 2018. p. 125–34.

68. Ezekowitz JA, O'Meara E, McDonald MA, Abrams H, Chan M, Ducharme A, et al. 2017 comprehensive update of the Canadian Cardiovascular Society guidelines for the management of heart failure. Can J Cardiol. 2017;33(11):1342–433.

69. Howlett JG, Chan M, Ezekowitz JA, Harkness K, Heckman GA, Kouz S, et al. The Canadian Cardiovascular Society heart failure companion: bridging guidelines to your practice. Can J Cardiol. 2016;32(3):296–310.

70. Fordyce CB, Douglas PS. Chapter 15 – Putting it all together: which test for which patient? In: de Lemos JA, Omland T, editors. Chronic coronary artery disease. Philadelphia: Elsevier; 2018. p. 204–25.

71. Blankenberg S, Zeller T. Chapter 9 – Standard and novel biomarkers. In: de Lemos JA, Omland T, editors. Chronic coronary artery disease. Philadelphia: Elsevier; 2018. p. 98–113.

72. Kwon L, Rosendorff C. Chapter 20 – The medical treatment of stable angina. In: de Lemos JA, Omland T, editors. Chronic coronary artery disease. Philadelphia: Elsevier; 2018. p. 280–302.

73. Tang J, Shen D, Caranasos TG, Wang Z, Vandergriff AC, Allen TA, et al. Therapeutic microparticles functionalized with biomimetic cardiac stem cell membranes and secretome. Nat Commun. 2017;8:13724.

74. Swanson N, Hogrefe K, Javed Q, Malik N, Gershlick AH. Vascular endothelial growth factor (VEGF)-eluting stents: in vivo effects on thrombosis, endothelialization and intimal hyperplasia. J Invasive Cardiol. 2003;15(12):688.
75. Yang J, Zeng Y, Zhang C, Chen Y, Yang Z, Li Y, et al. The prevention of restenosis in vivo with a VEGF gene and paclitaxel co-eluting stent. Biomaterials. 2013;34(6):1635–43.
76. Fihn SD, Gardin JM, Abrams J, Berra K, Blankenship JC, Dallas AP, et al. 2012 ACCF/AHA/ACP/AATS/PCNA/SCAI/STS guideline for the diagnosis and management of patients with stable ischemic heart disease: a report of the American College of Cardiology Foundation/American Heart Association Task Force on practice guidelines, and the American College of Physicians, American Association for Thoracic Surgery, Preventive Cardiovascular Nurses Association, Society for Cardiovascular Angiography and Interventions, and Society of Thoracic Surgeons. J Am Coll Cardiol. 2012;60(24):e44–e164.
77. Firouzi A. Chapter 20 – Percutaneous coronary intervention. In: Maleki M, Alizadehasl A, Haghjoo M, editors. Practical cardiology. St. Louis: Elsevier; 2018. p. 329–35.
78. Michos ED, Blaha MJ, Martin SS, Blumenthal RS. Chapter 29 – Screening for atherosclerotic cardiovascular disease in asymptomatic individuals. In: de Lemos JA, Omland T, editors. Chronic coronary artery disease. Philadelphia: Elsevier; 2018. p. 459–78.
79. Watson KE, Guo Y, Sahni S. Chapter 30 – Secondary prevention of coronary artery disease. In: de Lemos JA, Omland T, editors. Chronic coronary artery disease. Philadelphia: Elsevier; 2018. p. 479–87.
80. O'Gara PT, Kushner FG, Ascheim DD, Casey J, Donald E, Chung MK, de Lemos JA, et al. 2013 ACCF/AHA guideline for the management of ST-elevation myocardial infarction: a report of the American College of Cardiology Foundation/American Heart Association Task Force on practice guidelines. J Am Coll Cardiol. 2013;61(4):e78.
81. Ducrocq G, Steg PG. Chapter 21 – Antiplatelet and anticoagulant drugs. In: de Lemos JA, Omland T, editors. Chronic coronary artery disease. Philadelphia: Elsevier; 2018. p. 303–20.
82. Levine GN, Bates ER, Bittl JA, Brindis RG, Fihn SD, Fleisher LA, et al. 2016 ACC/AHA guideline focused update on duration of dual antiplatelet therapy in patients with coronary artery disease. JACC. 2016;68(10):1082–115.
83. Frampton J, Devries JT, Welch TD, Gersh BJ. Modern management of ST-segment elevation myocardial infarction. Curr Probl Cardiol. 2018;45(3):100393.
84. Capodanno D, Angiolillo DJ. Antiplatelet therapy after implantation of bioresorbable vascular scaffolds. J Am Coll Cardiol Intv. 2017;10(5):425.
85. Marso SP. Chapter 23 – Revascularization approaches. In: de Lemos JA, Omland T, editors. Chronic coronary artery disease. Philadelphia: Elsevier; 2018. p. 337–54.
86. Slicker K, Lane WG, Oyetayo OO, Copeland LA, Stock EM, Michel JB, et al. Daily cardiac catheterization procedural volume and complications at an academic medical center. Cardiovasc Diagn Ther. 2016;6(5):446–52.
87. Roubin GS. Author. The first balloon-expandable coronary stent: an expedition that changed cardiovascular medicine: a memoir [electronic resource]. St Lucia: University of Queensland Press; 2014.
88. Barton M, Grüntzig J, Husmann M, Rösch J. Balloon angioplasty – the legacy of Andreas Grüntzig, M.D. (1939–1985). Front Cardiovasc Med. 2014;1:15.
89. Singh IM, Holmes DR. Myocardial revascularization by percutaneous coronary intervention: past, present, and the future. Curr Probl Cardiol. 2011;36(10):375–401.
90. Byrne RA, Stone GW, Ormiston J, Kastrati A. Coronary balloon angioplasty, stents, and scaffolds. Lancet. 2017;390(10096):781–92.
91. Garg S, Serruys PW. Coronary stents: current status. J Am Coll Cardiol. 2010;56(10):S1.
92. Palmerini T, Biondi-Zoccai G, Della Riva D, Mariani A, Genereux P, Branzi A, et al. Stent thrombosis with drug-eluting stents. J Am Coll Cardiol. 2013;62(21):1915–21.
93. Watts TE, Chatterjee A, Leesar MA. Chapter 15 – Stent thrombosis: early, late, and very late. In: Topaz O, editor. Cardiovascular thrombus. San Diego: Academic Press; 2018. p. 217–24.

94. Shreenivas SS, Sarembock IJ, Kereiakes DJ. Chapter 16 – Stent thrombosis: implications for new stent designs and dual antiplatelet therapy duration. In: Topaz O, editor. Cardiovascular thrombus. San Diego: Academic Press; 2018. p. 225–47.

95. Dubel GJ. Angioplasty balloons, stents, and endografts. Tech Vasc Interv Radiol. 2000;3(4):214–25.

96. Mishra S, Bahl VK. Coronary hardware part 3 – balloon angioplasty catheters. Indian Heart J. 2010;62(4):335–41.

97. Garramone S. Structure-property relationships in angioplasty balloons. Worcester Polytechnic Institute, Digital WPI; 2001. https://web.wpi.edu/Pubs/ETD/Available/etd-0430101-122300/unrestricted/garramone.pdf

98. Park S, Bearinger JP, Lautenschlager EP, Castner DG, Healy KE. Surface modification of poly(ethylene terephthalate) angioplasty balloons with a hydrophilic poly(acrylamide-co-ethylene glycol) interpenetrating polymer network coating. J Biomed Mater Res. 2000;53(5):568–76.

99. U.S. Department of Health and Human Services, U. S. Food and Drug Administration. FDA executive summary circulatory system devices panel meeting: paclitaxel-coated drug coated balloon and drug-eluting stent late mortality panel; 2019.

100. U.S. Department of Health and Human Services, U.S. Food and Drug Administration. Premarket approval (PMA): IN PACT admiral paclitaxel-eluting percutaneous transluminal angioplasty balloon catheter; 2015.

101. U.S. Department of Health and Human Services, U.S. Food and Drug Administration. Premarket approval (PMA): Stellarex 0.035 OTW drug-coated angioplasty balloon; 2017.

102. U.S. Department of Health and Human Services, U.S. Food and Drug Administration. Premarket approval (PMA): Lutonix 035 drug coated balloon PTA catheter; 2017.

103. Hsieh M, Huang Y, Yeh J, Chen C, Chen D, Yang C, et al. Predictors of long-term outcomes after drug-eluting balloon angioplasty for bare-metal stent restenosis. Heart Lung Circ. 2018;27(5):588–94.

104. Ansel GM, Silver M, Phillips J, Jolly M. Coronary drug-coated balloons. Washington, DC: American College of Cardiology; 2016.

105. Liu L, Liu B, Ren J, Hui G, Qi C, Wang J. Comparison of drug-eluting balloon versus drug-eluting stent for treatment of coronary artery disease: a meta-analysis of randomized controlled trials. BMC Cardiovasc Disord. 2018;18:46.

106. Rittger H, Brachmann J, Sinha A, Waliszewski M, Ohlow M, Brugger A, et al. A randomized, multicenter, single-blinded trial comparing paclitaxel-coated balloon angioplasty with plain balloon angioplasty in drug-eluting stent restenosis: the PEPCAD-DES study. J Am Coll Cardiol. 2012;59(15):1377–82.

107. Li Y, Tellez A, Rousselle SD, Dillon KN, Garza JA, Barry C, et al. Biological effect on drug distribution and vascular healing via paclitaxel-coated balloon technology in drug eluting stent restenosis swine model. Catheter Cardiovasc Interv. 2016;88(1):89–98.

108. Meneguz-Moreno R, Ribamar Costa J, Abizaid A. Drug-coated balloons: Hope or hot air: update on the role of coronary DCB. Curr Cardiol Rep. 2018;20(10):100.

109. Task Fm, Windecker S, Kolh P, Alfonso F, Collet J, Cremer J, et al. 2014 ESC/EACTS guidelines on myocardial revascularization the task force on myocardial revascularization of the European Society of Cardiology (ESC) and The European Association for Cardio-Thoracic Surgery (EACTS) developed with the special contribution of the European Association of Percutaneous Cardiovascular Interventions (EAPCI). Eur Heart J. 2014;35(37):2541–619.

110. Sigwart U, Puel J, Mirkovitch V, Joffre F, Kappenberger L. Intravascular stents to prevent occlusion and restenosis after transluminal angioplasty. N Engl J Med. 1987;316(12):701.

111. Pache JÜ, Kastrati A, Mehilli J, Schühlen H, Dotzer F, Hausleiter J, et al. Intracoronary stenting and angiographic results: strut thickness effect on restenosis outcome (ISAR-STEREO-2) trial. J Am Coll Cardiol. 2003;41(8):1283–8.

112. Stefanini GG, Taniwaki M, Windecker S. Coronary stents: novel developments. Heart. 2014;100(13):1051–61.

113. Hanawa T. Materials for metallic stents. J Artif Organs. 2009;12(2):73–9.
114. Saraf AR, Yadav SP. Chapter 2 – Fundamentals of bare-metal stents. In: Wall JG, Podbielska H, Wawrzyńska M, editors. Functionalised cardiovascular stents. Cambridge: Woodhead Publishing; 2018. p. 27–44.
115. Hermawan H. Biodegradable metals electronic resource: from concept to applications. Heidelberg/New York: Springer; 2012.
116. Purnama A, Hermawan H, Mantovani D. Biodegradable metal stents: a focused review on materials and clinical studies. J Biomater Tissue Eng. 2014 Nov; 4(11):868–74(7).
117. Schewe S, Glocker DA, Ranade SV. Chapter 4 – Coatings for radiopacity. In: Medical coatings and deposition technologies. Hoboken: Wiley; 2016. p. 115–30.
118. O'Brien B, Carroll W. The evolution of cardiovascular stent materials and surfaces in response to clinical drivers: a review. Acta Biomater. 2009;5(4):945–58.
119. Trevor S, Benjamin H, Daniel RF, Michael F, Yong-Xiang C, Edward O. The evolution of coronary stents: a brief review. Can J Cardiol. 2014;30(1):35–45.
120. Khan W, Farah S, Domb AJ. Drug eluting stents: developments and current status. J Control Release. 2012;161(2):703–12.
121. U.S. Department of Health and Human Services, U. S. Food and Drug Administration. Establishment registration & device listing; 2019.
122. Neumann F, Sousa-Uva M, Ahlsson A, Alfonso F, Banning AP, Benedetto U, et al. ESC/EACTS guidelines on myocardial revascularization. Eur Heart J. 2018;2018:ehy394.
123. Bønaa KH, Mannsverk J, Wiseth R, Aaberge L, Myreng Y, Nygård O, et al. Drug-eluting or bare-metal stents for coronary artery disease. N Engl J Med. 2016;375(13):1242–52.
124. Zheng F, Xing S, Gong Z, Xing Q. Five-year outcomes for first generation drug-eluting stents versus bare-metal stents in patients with ST-segment elevation myocardial infarction: a meta-analysis of randomised controlled trials. Heart Lung Circ. 2014 Jun;23(6):542–8.
125. U.S. Department of Health and Human Services, U. S. Food and Drug Administration. Premarket approval (PMA): Resolute onyx zotarolimus-eluting coronary stent system; 2017.
126. Whitbeck MG, Applegate RJ. Second generation drug-eluting stents: a review of the everolimus-eluting platform. Clin Med Insights Cardiol. 2013;7:115–26.
127. Health, Center for Devices and Radiological. Recently-approved devices – EluNIR® ridaforolimus eluting coronary stent system – P170008; 2017.
128. Partida RA, Yeh RW. Contemporary drug-eluting stent platforms. Interv Cardiol Clin. 2016;5(3):331–47.
129. Alahmar A, Gershlick A. Drug-eluting stents—issues and developments. US Cardiol Rev. 2009;6(2):87–96.
130. Park DS, Bae I, Jeong MH, Lim K, Hong YJ, Shim JW, et al. Anti-restenotic and anti-thrombotic effect of polymer-free N-TiO2 film-based tacrolimus-eluting stent in a porcine model. Mater Today Commun. 2020;22:100777.
131. Lüscher TF, Steffel J, Eberli FR, Joner M, Nakazawa G, Tanner FC, et al. Drug-eluting stent and coronary thrombosis. Circulation. 2007;115(8):1051.
132. Wessely R, Schömig A, Kastrati A. Sirolimus and paclitaxel on polymer-based drug-eluting stents: similar but different. J Am Coll Cardiol. 2006;47(4):708–14.
133. Giordano A, Romano S, Monaco M, Sorrentino A, Corcione N, Di Pace AL, et al. Differential effect of atorvastatin and tacrolimus on proliferation of vascular smooth muscle and endothelial cells. Am J Phys Heart Circ Phys. 2012;302(1):H135–42.
134. Zhu S, Viswambharan H, Gajanayake T, Ming X, Yang Z. Sirolimus increases tissue factor expression but not activity in cultured human vascular smooth muscle cells. BMC Cardiovasc Disord. 2005;5(1):22.
135. Cho Y, Yang H, Park K, Chung W, Choi D, Seo W, et al. Paclitaxel- versus sirolimus-eluting stents for treatment of ST-segment elevation myocardial infarction. J Am Coll Cardiol Intv. 2010;3(5):498.

136. Axel DI, Kunert W, Goggelmann C, Oberhoff M, Herdeg C, Kuttner A, et al. Paclitaxel inhibits arterial smooth muscle cell proliferation and migration in vitro and in vivo using local drug delivery. Circulation. 1997;96(2):636–45.
137. Busch R, Strohbach A, Peterson S, Sternberg K, Felix S. Parameters of endothelial function are dependent on polymeric surface material. Biomed Tech (Berl). 2013; 58(1). https://doi.org/10.1515/bmt-2013-4053
138. Busch R, Strohbach A, Rethfeldt S, Walz S, Busch M, Petersen S, et al. New stent surface materials: the impact of polymer-dependent interactions of human endothelial cells, smooth muscle cells, and platelets. Acta Biomater. 2014;10(2):688–700.
139. Akinapelli A, Chen JP, Roy K, Donnelly J, Dawkins K, Huibregtse B, et al. Current state of bioabsorbable polymer-coated drug-eluting stents. Curr Cardiol Rev. 2016;13(2):139–54.
140. Ranade SV, Udipi Kishore, Glocker DA. Chapter 3 – Drug delivery coatings for coronary stents. In: Medical coatings and deposition technologies. Hoboken: Wiley; 2016. p. 75–114.
141. Kang S, Park KW, Kang D, Lim W, Park KT, Han J, et al. Biodegradable-polymer drug-eluting stents vs. bare metal stents vs. durable-polymer drug-eluting stents: a systematic review and Bayesian approach network meta-analysis. Eur Heart J. 2014;35(17):1147–58.
142. Han Y, Jing Q, Xu B, Yang L, Liu H, Shang X, et al. Safety and efficacy of biodegradable polymer-coated sirolimus-eluting stents in "real-world" practice. J Am Coll Cardiol Intv. 2009;2(4):303.
143. van der Heijden KMM, Zocca P, GAJ J, Schotborgh CE, Roguin A, et al. Bioresorbable polymer-coated orsiro versus durable polymer-coated resolute onyx stents (BIONYX): rationale and design of the randomized TWENTE IV multicenter trial. Am Heart J. 2018;198:25–32.
144. Lam MK, Sen H, Tandjung K, van Houwelingen KG, de Vries AG, Danse PW, et al. Comparison of 3 biodegradable polymer and durable polymer-based drug-eluting stents in all-comers (BIO-RESORT): rationale and study design of the randomized TWENTE III multicenter trial. Am Heart J. 2014;167(4):445–51.
145. Garg S, Serruys PW. Coronary stents looking forward. J Am Coll Cardiol. 2010;56(10):S43–78.
146. Grube E, Schofer J, Hauptmann KE, Nickenig G, Curzen N, Allocco DJ, et al. A novel paclitaxel-eluting stent with an ultrathin abluminal biodegradable polymer: 9-month outcomes with the JACTAX HD stent. J Am Coll Cardiol Intv. 2010;3(4):431–8.
147. Nogic J, McCormick LM, Francis R, Nerlekar N, Jaworski C, West NEJ, et al. Novel bioabsorbable polymer and polymer-free metallic drug-eluting stents. J Cardiol. 2018;71(5):435–43.
148. Baquet M, Jochheim D, Mehilli J. Polymer-free drug-eluting stents for coronary artery disease. J Interv Cardiol. 2018;31(3):330–7.
149. Kufner S, Sorges J, Mehilli J, Cassese S, Repp J, Wiebe J, et al. Randomized trial of polymer-free sirolimus- and probucol-eluting stents versus durable polymer zotarolimus-eluting stents: 5-year results of the ISAR-TEST-5 trial. J Am Coll Cardiol Intv. 2016;9(8):784–92.
150. Lee JH, Kim ED, Jun EJ, Yoo HS, Lee JW. Analysis of trends and prospects regarding stents for human blood vessels. Biomater Res. 2018;22:8.
151. De Luca G, Smits P, Hofma SH, Di Lorenzo E, Vlachojannis GJ, Van't Hof AWJ, Arnoud WJ, et al. Everolimus eluting stent vs first generation drug-eluting stent in primary angioplasty: a pooled patient-level meta-analysis of randomized trials. Int J Cardiol. 2017;244:121–7.
152. Kawakami R, Hao H, Imanaka T, Shibuya M, Ueda Y, Tsujimoto M, et al. Initial pathological responses of second-generation everolimus-eluting stents implantation in Japanese coronary arteries: comparison with first-generation sirolimus-eluting stents. J Cardiol. 2018;71(5):452–7.
153. Byrne RA, Joner M, Kastrati A. Stent thrombosis and restenosis: what have we learned and where are we going? The Andreas Gruntzig lecture ESC 2014. Eur Heart J. 2015;36(47):3320.
154. Buchanan K, Steinvil A, Waksman R. Does the new generation of drug-eluting stents render bare metal stents obsolete? Cardiovasc Revasc Med. 2017;18(6):456–61.
155. Philip F, Agarwal S, Bunte MC, Goel SS, Tuzcu EM, Ellis S, et al. Stent thrombosis with second-generation drug-eluting stents compared with bare-metal stents: network meta-analysis of primary percutaneous coronary intervention trials in ST-segment–elevation myocardial infarction. Circ Cardiovasc Interv. 2014;7(1):49–61.

156. Tada T, Byrne RA, Simunovic I, King LA, Cassese S, Joner M, et al. Risk of stent thrombosis among bare-metal stents, first-generation drug-eluting stents, and second-generation drug-eluting stents: results from a registry of 18,334 patients. J Am Coll Cardiol Intv. 2013;6(12):1267–74.
157. Navarese EP, Kowalewski M, Kandzari D, Lansky A, Górny B, Kołtowski Ł, et al. First-generation versus second-generation drug-eluting stents in current clinical practice: updated evidence from a comprehensive meta-analysis of randomised clinical trials comprising 31 379 patients. Open Heart. 2014;1(1):e000064.
158. Brancati MF, Burzotta F, Trani C, Leonzi O, Cuccia C, Crea F. Coronary stents and vascular response to implantation: literature review. Pragmat Obs Res. 2017;8:137–48.
159. Otsuka F, Finn AV, Yazdani SK, Nakano M, Kolodgie FD, Virmani R. The importance of the endothelium in atherothrombosis and coronary stenting. Nat Rev Cardiol. 2012;9(8):439–53.
160. Otsuka F, Byrne RA, Yahagi K, Mori H, Ladich E, Fowler DR, et al. Neoatherosclerosis: overview of histopathologic findings and implications for intravascular imaging assessment. Eur Heart J. 2015;36(32):2147–59.
161. Inoue T, Node K. Molecular basis of restenosis and novel issues of drug-eluting stents. Circ J. 2009 Apr;73(4):615–21.
162. Fukuda D, Sata M, Tanaka K, Nagai R. Potent inhibitory effect of sirolimus on circulating vascular progenitor cells. Circulation. 2005;111(7):926–31.
163. Meyers SR, Kenan DJ, Khoo X, Grinstaff MW. A bioactive stent surface coating that promotes endothelialization while preventing platelet adhesion. Biomacromolecules. 2011;12(3):533–9.
164. Butzal M, Loges S, Schweizer M, Fischer U, Gehling UM, Hossfeld DK, et al. Rapamycin inhibits proliferation and differentiation of human endothelial progenitor cells in vitro. Exp Cell Res. 2004 Oct 15;300(1):65–71.
165. Im SH, Jung Y, Kim SH. Current status and future direction of biodegradable metallic and polymeric vascular scaffolds for next-generation stents. Acta Biomater. 2017;60:3–22.
166. Ang HY, Bulluck H, Wong P, Venkatraman SS, Huang Y, Foin N. Bioresorbable stents: current and upcoming bioresorbable technologies. Int J Cardiol. 2017;228:931–9.
167. Dave B. Bioresorbable scaffolds: current evidences in the treatment of coronary artery disease. J Clin Diagn Res. 2016 Oct;10(10):OE01–7.
168. Borhani S, Hassanajili S, Ahmadi Tafti SH, Rabbani S. Cardiovascular stents: overview, evolution, and next generation. Prog Biomater. 2018;7(3):175–205.
169. Du F, Zhou J. Vascular intervention: from angioplasty to bioresorbable vascular scaffold. In: Fu BM, Wright NT, editors. Molecular, cellular, and tissue engineering of the vascular system. Cham: Springer; 2018. p. 181–9.
170. Hideo-Kajita A, Wopperer S, Seleme VB, Ribeiro MH, Campos CM. The development of magnesium-based resorbable and iron-based biocorrodible metal scaffold technology and biomedical applications in coronary artery disease patients. Appl Sci (Basel). 2019 Sep 1;9(17):3527.
171. Inoue T, Croce K, Morooka T, Sakuma M, Node K, Simon DI. Vascular inflammation and repair: implications for reendothelialization, restenosis, and stent thrombosis. JACC Cardiovasc Interv. 2011;4(10):1057–66.
172. McGonigle J, Webster TJ, Bhardwaj G, Glocker DA, Ranade SV. Chapter 5 – Biocompatibility and medical device coatings. In: Medical coatings and deposition technologies. Hoboken: Wiley; 2016. p. 131–80.
173. Wu D, Yu M, Gao H, Zhang L, Song F, Zhang X, et al. Polymer-free versus permanent polymer drug eluting stents in coronary artery disease: a meta-analysis of 10 RCTs with 6575 patients. Chronic Dis Transl Med. 2015;1(4):221–30.
174. Richards CN, Schneider PA. Will mesh-covered stents help reduce stroke associated with carotid stent-angioplasty? Semin Vasc Surg. 2017;30(1):25–30.
175. Park KY, Kim DI, Kim BM, Nam HS, Kim YD, Heo JH, et al. Incidence of embolism associated with carotid artery stenting: open-cell versus closed-cell stents. J Neurosurg. 2013;119(3):642–7.

176. Yang Z, Tu Q, Wang J, Huang N. The role of heparin binding surfaces in the direction of endothelial and smooth muscle cell fate and re-endothelialization. Biomaterials. 2012;33(28):6615–25.
177. Capodanno D, Dipasqua F, Tamburino C. Novel drug-eluting stents in the treatment of de novo coronary lesions. Vasc Health Risk Manag. 2011;7:103–18.
178. Poh CK, Shi Z, Lim TY, Neoh KG, Wang W. The effect of VEGF functionalization of titanium on endothelial cells in vitro. Biomaterials. 2010;31(7):1578–85.
179. Tang C, Wang G, Wu X, MS LZ, MD SY, Lee JC, et al. The impact of vascular endothelial growth factor-transfected human endothelial cells on endothelialization and restenosis of stainless steel stents. J Vasc Surg. 2011;53(2):461–71.
180. Liang C, Hu Y, Wang H, Xia D, Li Q, Zhang J, et al. Biomimetic cardiovascular stents for in vivo re-endothelialization. Biomaterials. 2016;103:170–82.
181. Yang Z, Yang Y, Xiong K, Li X, Qi P, Tu Q, et al. Nitric oxide producing coating mimicking endothelium function for multifunctional vascular stents. Biomaterials. 2015;63:80–92.
182. Bedair TM, ElNaggar MA, Joung YK, Han DK. Recent advances to accelerate re-endothelialization for vascular stents. J Tissue Eng. 2017;8:2041731417731546.
183. Ye C, Wang Y, Su H, Yang P, Huang N, Maitz MF, et al. Construction of a fucoidan/laminin functional multilayer to direction vascular cell fate and promotion hemocompatibility. Mater Sci Eng C. 2016;64:236–42.
184. Zhong S, Luo R, Wang X, Tang L, Wu J, Wang J, et al. Effects of polydopamine functionalized titanium dioxide nanotubes on endothelial cell and smooth muscle cell. Colloids Surf B: Biointerfaces. 2014;116:553–60.
185. Chang H, Kim P, Kim DW, Hyun-Min C, Mi Jin J, Dea Han K, et al. Coronary stents with inducible VEGF/HGF-secreting UCB-MSCs reduced restenosis and increased re-endothelialization in a swine model. Exp Mol Med. 2018;50(9):1–14.
186. Singh A, Singh A, Sen D. Mesenchymal stem cells in cardiac regeneration: a detailed progress report of the last 6 years (2010–2015). Stem Cell Res Ther. 2016;7(1):82.
187. McKittrick CM, Cardona MJ, Black RA, McCormick C. Development of a bioactive polymeric drug eluting coronary stent coating using electrospraying. Ann Biomed Eng. 2020;48(1):271–81.

Index

Printed in the United States
By Bookmasters